In the Mirror of Nature

The Unspoken Language of Our Interconnectedness with Earth

Kristen Winter, Ph.D.

The Awakened Press
www.theawakenedpress.com

For additional information, please contact The Awakened Press at books@theawakenedpress.com.

The Awakened Press can bring authors to your live event. For more information or to book an event, contact books@theawakenedpress.com or visit our website at www.theawakenedpress.com.

Front Cover Photo of Crow and Author Biography Photo: © Kristen Winter, Ph.D.

Editing and Book Development: Lindsay R.A. Dierking, Angela Marie
Cover and Book Design: Kurt A. Dierking II

Printed in the United States of America
Distributed throughout the United States of America and worldwide
First The Awakened Press edition

Paperback ISBN: 979-8-234-10114-3

For anyone who may sense a lot from their surroundings.
For those who might feel alone in a world that can seem isolating and confusing.

For our uniquely beautiful, sacred Earth and all of her supportive
communities—the visually stunning, emotionally vibrant, and brilliantly
intelligent individuals who matter as they are, profoundly.

For all the individual lives throughout Nature, close to home and sharing our neighborhoods.

For those individuals who have touched and changed my life in beautifully
poignant, often subtle moments—these most meaningful links of connection.

For Willow.

Table of Contents

Earth Speaks Softly

My inspiration has come from individual life throughout our natural world. It was not always readily apparent, even to myself, but there has always been something within me—almost a need—to do something that makes a difference. Not through big adventures or far-off places, but through the beauty of community close to home. Places just outside cities, some more urban than not, others quieter and more secluded. Moments of connection on my front porch, balcony, in my yard, and in the places where I have worked across many moves.

The most special, intelligent, and inspiring support has come from the lives within our neighborhoods and communities. I have learned from them and with them, often through very personal moments that felt meant only for me, always in quiet calm. These relationships have changed me forever.

The most profound moments in life do not always come from other people or from situations we create or feel we need. They happen all around us, daily, with life in Nature, and we can learn from them. We learn that these lives are individuals just like us—individuals who play when safe and suffer through pain and loss. We find it easier to feel when we allow ourselves this path of growth.

These lives matter to us because our health is suffering without them. They matter simply because they matter. Through them, and often in ways that go unnoticed, we are changed—within body, mind, inspiration, and empowerment.

In these written words, I am sharing what I have learned from life in Nature. In my desire to understand human relationships, I came to realize that our relationships with life in Nature guide us because of who we are as individuals. In flashing moments or quiet expressions over time, Nature reveals awareness. Through astonishing brilliance, these lives show us what relationships can mean in terms of authentic trust, emotional strength, and the healing of our interconnected Earth—all stemming from our sense of self. As with any relationship, it is never one-sided.

Learning to Listen

Something within me awakened when I was very young. It began with one individual Squirrel I wanted to know well, an accidental witnessing of Animals suffering in fear and pain, and an adult telling me that Dogs and all life in Nature do not matter. I kept much of myself quiet and wanted deeply to feel less alone. As I grew older, I wanted to understand trust and respect and how they impact relationships, regardless of who another is or what form they take.

I felt torn between wanting to help children and wanting to help the environment. I earned a B.S. in psychology from a traditional but open-minded university. Later, I withdrew from a graduate program in art therapy to begin working as an environmental educator and volunteering with conservation organizations, beginning a proud career in nonprofit environmental conservation. The work tested me at times, but I felt called toward mission-driven environments where I could contribute alongside others who felt the same drive.

Early on, I searched for a way to combine psychology and ecology. Around 1995, I discovered a book on ecopsychology and nature-focused therapy and felt thrilled by what I found, though there were no suitable programs available to me at the time. Eventually, I earned an M.S. in environmental protection and safety management while working both full-time and part-time in local nature centers and with international organizations. I wanted to understand environmental responsibility closely. Yet I still struggled to understand why so many people seemed so emotionally removed from other life on Earth.

Willow and the Empowerment of Belonging

I began making bold changes where change felt necessary, learning more about myself and others along the way. I adopted Willow, my Dog and soulmate, who became my constant companion through moves, challenges, and transitions. Later, I enrolled in a remote Ph.D. program in applied ecopsychology—exactly the path I had been searching for. I felt at home in the work and empowered by it, writing rapidly and eagerly. I completed my dissertation and earned my Ph.D. in 2017, with Willow beside me throughout the journey.

I continue learning and growing, knowing now that our health and happiness depend upon interconnectedness with life in Nature, especially close to home, where we need inspiration most.

These lives matter in ways we have yet to fully understand or appreciate. They matter on their own terms and through their own worth. With them, and through them, we can thrive more considerately and more fully.

After seventeen years and three months, Willow is no longer in pain. Yet her brilliance remains a grounding strength within me—a reminder of loyalty to personal truth and to life throughout Nature. Regardless of place, we are interconnected. On this incredible Earth that belongs to all of us together, we feel better and become better. Here breathes our empowerment. It begins within and with each one of us.

Why This Book Exists: A Message Through Nature

With a personal and focused approach to individual life in Nature, this book began as an expansion of my dissertation. As I found support and understanding within the natural world, I slowly realized I had become part of a larger message to humanity. My hope is that others come to know Nature in ways that inspire and empower them through these relationships. I have learned from emotionally and mentally intelligent individuals all around us in the natural world. Relationships formed in moments and strengthened across time with Nature have become as necessary and trusted to me as breath itself.

This work is intended to create a flow that begins with the struggles many people experience. In many ways, these opening sections reflect life before the connections I longed to understand within myself. They explore the moments that cause many of us to pause and feel lost. While some sections may feel heavy, they are meant to focus honestly on issues that remain deeply alive within our world.

From there, the experiences begin building through life with Nature in the chapters that follow. These lives, as I have come to know them, are our support. They reveal how and why we were never meant to become so emotionally separated from the natural world. The content reaches deeper into why I believe this to be true and includes many of the experiences that shaped this understanding. I attempt to share how my own life changed because of what life outside my door taught me.

There is also space for the reality that we fall back into old patterns and ruts, including those created simply by living through personal struggles and experiences. Ultimately, however, these struggles can become strengths. My hope is that, by the end, it becomes clear why our relationships with Nature matter so deeply. As part of Nature ourselves, we carry this interconnectedness throughout our lives.

What follows is what I have learned through vital, uniquely profound, and considerate relationships with individuals from our brilliant natural world. With Nature held within us, and all of us together on this sacred Earth, we are never alone. Within, around, and through us, life quietly reveals why our relationships with Nature matter.

Part One

Isolated from Nature Within

One
Reality Ignored

Nature Is Dismissed

*T**oo much, too fast, and too available* is making us ill. There is no time to *be*. No time for meaning. No time for what is real.

We can feel lost in a world where everything is available at our fingertips. Why are we here? What can we do to make life better for ourselves and for the world? We've met many of the requirements for happiness, yet we tend to struggle to catch our breath. In rare moments of quiet, we're pulled to find our way.

Despite our accomplishments, many still feel not quite right. Loneliness settles in—sometimes suddenly, sometimes all the time. We have moments where we don't even know what we feel or how to change it. Some feel lonely without knowing why. This is a great contradiction in our lives… Something is missing.

We get hurt again and again, and we blame others for our pain. We self-medicate, numb out, consume too much, and create more imbalance while trying to fix the imbalance. We keep our heads down, eyes on our tasks, minds racing.

We place so much worth on so many things that do not matter at all once we are confronted with our own lives, health, and basic needs. We find endless time for screens, always searching and viewing another world that is not present and with us, and is very limited in what matters most in life.

Worth is often placed on the aggressive, the dominant, the loudest talker. We speak constantly, and yet we miss what's really going on. We rely on language to communicate, but we lie with words. We're confused, frustrated, unsure if we've even communicated anything real. We ignore other ways of knowing—ways that could heal us—because they are considered vulnerable or weak. But words often separate us from truth. We ignore our emotional intelligence—the very intelligence that connects us with all of life. This divides us from life in Nature.

Some of us—those seen as overly sensitive, deeply creative, and aware in ways that make others uncomfortable—have long been misunderstood. We isolate these parts of ourselves. We're uncomfortable feeling our own experience, let alone anyone else's.

Isolated from Nature within, our bodies feel lonely, too. As a people, we suffer mentally and physically. We feel exhausted, unclear, and without purpose. We struggle to find our center—something to motivate us, to support all the good we want to bring into the world. We've worked so hard and still feel unfulfilled.

We can't see our own light when we're blinded by the light of others we believe hold the answers. In those times, we feel like our light is dimmed or we're without light.

Constrained by systems that stifle us, negativity grows within us and around us. It floods our thoughts and cells. We poison ourselves with what we consume mentally and emotionally. Fear feeds anger. And trapped inside, this unresolved fear and anger fester. It makes us sick. It leaks into the world. We fear discomfort, so we grow comfortable with pain. The invasive weeds grow quickly, smothering any Flowers. Negative creates negative—fast. And we ignore the reality of Nature's pain, too.

We can't see inside our own cells, but we *feel* it in our bodies, in the mirror, in our bones. We fill the void with temporary solutions and ignore our health and happiness. Our mind-body connection is mostly ignored. In doing so, we lose our purpose.

We don't live in a vacuum. Our worlds are more than just our personal relationships. Yet many use fear and control disguised as "power" to hurt others. And when we act from fear and anger, the consequences ricochet back as animosity and emptiness.

We're taught that emotional connection with Nature is childish. We assume that the lives throughout Nature are not sentient beings with thoughts and minds, and that these lives act only on instinct. We dismiss one another's sensitivity as weakness and ignore Nature's excellence. We think we are the only ones with ideal emotions—but even we are numbing our own.

Some create drama to feel something—anything. Yet, we fear feelings. And in this fear, we stop knowing ourselves. We become more removed, more restricted, more alone. Disconnected, we turn on each other. Ultimately, we mistreat others, ourselves, and the living world around us.

We've created a world where Nature is expected to serve us and adapt to us. We readily brush off Animal and natural life experiences just because they express pain and joy differently. But their suffering—like our own—is real.

This false sense of superiority limits our ability to see others clearly. We justify doing whatever we want because we assume other lives are beneath us. But Insects, soil microbes, and entire ecosystems are vital to this planet. They are vital to each one of us. Without them, our world collapses.

Nature respects boundaries and will push back when pushed too far. When we stop listening, the imbalance grows.

We ignore Nature in our ethics, our relationships, and our systems—even though it wasn't always this way.

Wisdom from past generations has been erased or buried, passed down now only as struggle. Today, we don't know how to shift, and we feel this restriction. We're missing the signs and ignoring our own inner compass.

We want to control our environments, yet we're excluded from the deeper conversations of life. Nature works together. We, by contrast, starve our inner connection to this life, this Earth. We try to rip Nature out of ourselves, but it's a self-induced straitjacket. Nature is seen as "things" to conquer, buy, or use—not within us to include or belong with. By harming others, we echo this harm throughout all of Nature.

But Nature continues around us, whether we're listening or not. And until we do, we'll remain out of balance and haunted by an unidentifiable longing.

Quick fixes and instant gratification are not new. Many of us have been lost, lonely, and misdirected for a long time. We believe we need something outside ourselves to make us feel complete. Our mindset is off.

It's easier to make excuses than to sit with the discomfort of what we are doing. We see this in how we treat Nature.

Through control, extermination, and abuse, we ignore Nature's individuals. But we need Nature—deeply—in our communities, our health, and our sense of self. There is no external location that will heal us. Most don't even feel these common threads anymore. *The separation lives within us.* We have forgotten that Nature *must breathe through us.*

We generalize. We repeat old patterns. We keep emotional distance. We describe Nature as objects—again, "things" to be discarded or feared. *We don't know Nature because we don't know ourselves.*

The roots of life mirror the roots in us. And we tell ourselves it doesn't matter—just as we do with our own needs. We mistreat life because we mistreat ourselves. *We have stopped believing that we are Nature, too.*

What we do to Nature returns to us.

We've forgotten we are part of a larger, living community that also wants us to thrive as individuals.

Our wellness will be reflected in Nature.

Even with good intentions, captivity of the natural world hurts. When lands are gone, communities are broken, and individuals are "protected" in cages longing for home, for their companions, and for their families—these broken relationships are also ours.

Why do we continue to believe that life throughout Nature is beneath us, unintelligent, and unworthy?

As a species, our hands cannot be trusted. These broken relationships are ours. If we are to change, we must go within.

We deny Nature a voice. We project our desires onto lives meant for freedom. They suffer—just as we would. And in doing so, we limit not only ourselves, but our evolution.

Somehow, still, we are never alone.

While our relationships with people are often transactional—our inner worlds dominating ourselves—individual lives in Nature can draw us into an experience with color and life. Nature connects us more vibrantly to our inner Nature, and to our physiology, profoundly and with so much life and color that is meaningful. With Nature, there is no isolation within.

The Forgotten Link

Our thoughts create reactions and changes in our bodies. A world of loneliness closes in when our means of connection go against all that is real. The functions of our systems require health and positivity to work properly. Just as our experiences create either supportive or toxic chemicals in our brains and bodies, our experiences around us are linked within. Many of us feel this loss in our lives, no matter how connected we may appear with our devices. Our neglected, broader intelligence keeps our foot tethered from stepping forward.

Regardless of how Nature appears to us, life in the natural world survives on feelings and all that is sensed. Insects share, teach, learn, and offer support in ways we may have never considered. As with our experiences, emotions flood through our bodies—without a word spoken. Our emotions are vital and linked to memory, just like life in the natural world.

All our experiences either add to or take from us. As subtle or strong as our feelings are toward life in Nature, our inner driver—our motive and intent—directs right back to us, within us, and around us. In these absent connections, we dismiss ourselves. One is not complete without the other. Our bodies understand this. Our brains could use a bit of catching up. Without both, we feel it.

We are the ones who have neglected and shut down Spider in the corner, tending to the beauty and necessity of their web. Ignoring this intelligent information, communication, and knowledge that radiates from Nature, we starve ourselves at our core. Loss is not human alone. Some have witnessed this with other life in Nature. They feel and grieve in ways that, for some, become too deep in depression (as we define it) to return.

Tied to our senses from our experiences and interactions, our emotions are reactions that produce our thoughts. This connects us. With Nature, we can calm and breathe. Our thoughts impact our mental and physical functions through activity in the brain and from our actions. Our thoughts create chemicals. These changes within flow as information throughout the brain, body, and organs. Our mind—our path to consciousness—is much older and wiser.

With individual life in the natural world, these connections run deep. We are never without connection to life around us. The presence of our mind impacts both the world outside and within.

We are often known through our facades and unawareness of self. But our actions reflect through us—and they matter.

When we are mindfully present in real, extended moments with life in Nature—near our homes, with our potted Plants, and in our yards—it becomes nearly impossible to struggle when we connect with the strength and intelligence of life in a way that is comfortable and appropriate—for them. We find solace in stepping out of the racing, thought-filled mind to live in very real moments. We return to what matters. We motion to life around us the space to be. It might be surprising how easy and natural it is to be with and learn from life nearby—at home and in our neighborhoods. This is where we learn. It is immediate and lasting.

For the most part, our world revolves around spoken words. We need language, but we need more, too. These other ways of understanding—nonverbal communication, communication through our senses—are common to life on Earth. They echo from the eyes of Bee, the careful knowing of Bird, the grounding knowledge flowing from Tree. These are like a lightning-fast shout, and as quiet and as still as silence. This might be best understood as energy. Not ours, but those sensed. The language of interconnected life.

Outside on the porch, focusing on lone Tree, our thoughts are left neatly on the table. Here, we are triggered by what we sense through experience with another individual—immediately and lasting, still.

Nature is always trying to return to balance. Stability cannot exist when one does not allow another a voice of their own. The kind of control people radiate has no place in Nature. The insecurity that creates control is not present among this brilliance. Balance and control cannot live in the same home. Yet, balance and control thrive within boundaries. Unlike our need to grasp or shake the boundaries of others, personal boundaries have a vital purpose in all life forms. For life in Nature, balance and boundaries are everything.

This need for balance radiates through all life. Similarly, our balance starts within us. This is where help for Nature lives. Perhaps the control needs to turn inward since so much feels out of control with our own boundaries.

Our need for one another is rooted in the truth that we are natural beings, too. We live on this planet, too. With Nature, there is understanding and adaptation based on the past, but focused on the present and future. If they remain this way, then so can we. It is within us. Slowing down to be present with life is healing. Why not listen to Nature?

There are those with highly developed senses—those gifted empathetically and intuitively—who can perceive changes from daily experiences in various ways. These individuals are more sensitive to the emotions, feelings, and experiences of other living beings, including people. They see clearly through facades. Life in Nature is aware. Who we are—our motive and intent—affects our mental and physical state, and this is perceptible to Nature through smell or other senses we may have dulled through lack of use. No excuses. Nature senses this, too. In Nature, our true essence is revealed.

A unifying bond with life in Nature dissolves our belief that our worth is greater or more critical. Individuals in Nature experience life like we do—beyond beauty, beyond what is taken from them. They express needs, interact, rely on senses, and feel deep and unique emotions. This sharing of information travels through many senses, possibly some unknown. Perhaps those of us with intact empathic abilities and those with deeper intuitive understandings are more aligned with the natural world, making us more proficient in building these relationships. In turn, we become more united with what we feel through what is sensed and felt by other life around us.

I met a woman who once told me she loved living in the woods and, in the same breath, proudly declared that she would run over an individual Snake warming on the road. A complete oxymoron of appreciation—and nothing of love. These lives want nothing more than to be free of us, given our disregard for boundaries. There is a way to live *with* them, helping them get to where they want to be. What they seek is what we seek: kindness and understanding. Often it means leaving them alone.

Those of us who connect more readily with life in Nature, when we approach these relationships in a way that values and honors their uniqueness and importance, we are supported in being who we are, regardless of pressure to be something else. We are supported with Nature when we are supporting within.

We will understand the voices of Nature through what we sense and feel when sitting with them—quietly. Quietly, what we need aligns for our calm to balance. Life, thought-out from Nature, can stop us—halt thinking—and simply let us be...*if* we are accepting. Balance is Nature's work, easing restlessness and supporting our coexistence in health.

In our search, life in Nature watches us and senses quietly as we go in and out of our homes, cars, offices, and yards. Nature is not numb to pain and grief—certainly not to struggle, given all they must overcome just to live each day. Where are our roots to the present and to Nature? All life wants respectful relationships. Nature, too, affirms positively when we allow them the freedom to speak in their own way.

Squirrels hold a special place in my heart. I remember a moment in early childhood with one specific Squirrel who stopped along the phone line. Standing in the backyard, I looked up, and Squirrel looked back. I desperately wanted to understand Squirrel. I wanted us to know one another—not through words. I tried with my mind. I thought, "I want to communicate with you." I didn't know how, but it seemed possible—perhaps natural for Squirrel. I thought that wanting it was enough. Nothing came back. I wasn't even sure what I expected—a flash of thought? A feeling? An image? Something not of me but felt by me? I didn't know—but I wanted it. I wanted to start a conversation.

As a child and beyond, as I tried to find my way by relying on experience, I failed to acknowledge those I cared for deeply. They were waiting for me to see and know them. I had it in me to be open, but I didn't know what that meant. Despite my intentions and growing interest, I ignored Nature and our relationship. I never paused long enough to know their intelligent force. I didn't know there were more senses than the few we are taught. I had to learn to halt intellect and language to open the door to new ways of experiencing. And I had to heal within. Still, I felt them shifting within me.

Over time, I remained unaware of how to honor those other forms of knowing. I ignored a different way of being and communicating. I restricted reality.

Standing in silence, Squirrel and I looked at each other. Then, Squirrel went on their way. I recall a deep determination mixed with real sadness. I knew the impasse was on my end. I whispered under my breath, "Someday." Before we can experience what we want positively on this sacred Earth, we need to become it within.

Without life in Nature in our hearts, we remain locked in our brains. Our minds and bodies don't work this way. We are Nature, too. Life in Nature operates through elaborate senses, while we've relied mostly on elaborate words. All of us have powerful minds, but we've stirred them into frenzy when they seek calm. What is in our Nature? We feel it—even if disconnected, the spark remains. It will always seek a response until we acknowledge it. Until we become it. We need balance. We need our interconnectedness.

The Quiet Beneath the Noise

Struggling can make us feel so very alone at times—lost in a sea of people and in a life that can feel like too much, and in a way where we might not even be able to assign words. We dismiss Nature—Nature all around us, Nature within us. We dismiss all of the lives that are Nature, collectively. We dismiss ourselves. Yet, we are all the same in our different flaws. How we choose to treat ourselves and, therefore, others in reflection of our flaws determines our calm.

What is of great value to remember here is that there are events and situations in life that will shake us—and sometimes to our core. Life in Nature shares in these experiences that trigger the slow burn within. We are not alone in these states of pain and restlessness for relief; moments of feeling at complete loss and not wanting to even move physically. Sometimes these events are a build-up of situations. There are times when the struggle seems to begin outside of our sense of self, starting as a small fire that then ignites within after smoldering for years. This fire damages and, in moments within, can rage out of control. Whether we need people to support us or time alone, finding that place of calm with life in Nature is, was, and will always be our enduring ally. We are never alone with life in Nature. Opening our field of vision to include them helps us—both.

We can also cause Nature's mental decline through emotional abuse. I have watched a Dove stand over her baby who had been mauled to death by an outdoor Cat left alone in a world that is unsafe for both the family Cat who becomes ill (and exposes their humans to these illnesses, unwillingly) and those who fall prey, such as our rapidly dwindling Songbird family or Voles who form a single partner relationship for life. I have also witnessed a sweet Cat family attacked by Cats left unattended outside, and, because not part of the natural community, they became very ill and ill-tempered. Attacks are ruthless. Family pets can become victims.

As for this mother Dove, she would not leave her baby's side after their struggle to live. Her heartbreak was palpable. Depression is real, and although this Dove did not weep tears as we do, she was distraught in a way that was easily felt—even by another person I was with in this moment, who was not as in tune with others' emotions. For me, as a witness, to think back still brings the emotion she felt shooting back into my gut—a center of neurotransmitters like the brain. Dove may have expressed her need for love for her baby differently. No way does this indicate that she was not feeling horrible loss as a mother, having witnessed what she did, unable to do anything to help her little one's suffering.

Life in Nature is not spared from depression. Losing a loved one, a mate, a baby, a friend, or even a human companion creates pain, suffering, and intense grief. We are not alone in our struggles—never. Perhaps if we allow for life in Nature to experience what they actually feel and not belittle their emotions—belittling them and their care, treatment, and needs—our unrestricted minds might begin to heal, and heal us in the process.

So many of us have experiences—fleeting or perhaps more regularly—with life in the natural world that tell a very different story from what we have known. If we would only listen.

We are not limited to experiences with other people. If we would only receive the awareness. With broken links to those lives that are Nature, the drive within is not gone—it only searches in the quiet beneath our experiences.

It is surprising that much of this complex relationship we have with anyone nonhuman largely revolves around our physical containment of vocal cords—that we speak words and have created a system for our language. Even if a lot of us fail to use it properly and clearly, myself included. There are other ways to communicate and share information; perhaps ways that are much more reliable and readily apparent across all life on this precious planet we all share. Words are essential and crucial, but so is so much more.

In moments where we are skilled in confidence to stop and listen to the positive information and communication of life in Nature, we begin to *know* Nature. Perhaps we may hear from Tree in front of our home or at the corner in the park, or from a small, raised garden of Flowers we created in our tiny space, one potted Plant, sky, the air outside, and the breeze sending us a plethora of information from an array of far-off places.

Among so much more, I have learned that Insects honor trust and show compassion in their freedom to be. Different, yes, but different is necessary for our Earth to survive. Yet different, along with alone, only seem to be celebrations when we are truly comfortable in our own skins of who we are as individuals. *Different* is essential. And alone does not equal lonely.

The reality is, we are very much alike life throughout the natural world. Our deep connection to life throughout Nature is needed more than they do of us. In Nature, there is support for one another—throughout systems and across great distances. There is interconnectedness. Is it too late for these relationships of ours, struggling by our own hand? Somehow, through all of it—with all the poor treatment and damage and complete lack of respect we radiate and inflict—they are much more ready to extend a hand than we.

We may do well for those in our lives; we are grateful when our families are doing well. If we want to help the environment—help life in Nature—perhaps a basic focus-shift to: we need life in Nature in a different, more emotional way within our center, our minds, and our consciousness. With Nature, we need an honest look inward into who we are, why we are, and how we need help within, so that we might *be*. This is vital and will help us. Our honesty to self in the present is essential.

Mindfulness is a powerful tool. Our mind can heal our bodies, and our mind can hurt our bodies. With the natural world, we can learn of them—their needs, who and how they are—in any grounding hopes of developing a kinship. Here, we are not learning alone.

The ability to remain aware and conscious is an extraordinary path and opening to life. History, including our own, can be such a valuable tool to shape ourselves, our lives, our vision, and our goals, wrapped in our desire to move forward positively throughout our days. We carry history with us, but we make our futures each day. Accepting our lives in the present and going forward as a manifestation of our own making through our own control, regardless of what we have lived through and what has happened to us, can empower us through responsible thoughts and actions—especially when we shift our minds and thoughts to create good each day in interactions with others. We are responsible for our health. We are responsible for our actions, our thoughts, and our deeds. We make or break our happiness—no one else. This comes from our strength within.

With all our amazing advancements in science and our understanding of desires for finding meaning behind personal beliefs, many of us feel our worlds are anything but perfect. With Nature, we feel the shift. Purpose enters. Purpose is different.

Purpose extends beyond; farther. Integrity-driven purpose within communities supporting communities in mindful kindness and health and happiness is very real and has energy all its own—and all from our within. This is our truest connection to the natural world. Purpose is for the good of the whole, including Earth. Starting from our within, we are each responsible for ourselves as we radiate that which we are out into the world—it all comes back to us. This is the beauty of life, and where we can learn and grow in mental, mindful, emotional, and spiritual strength. And when our mental self feeds our emotional and mindful parts of self, supported by a healthy spiritual self if we so choose, alignment begins within. We can feel it. So can life throughout Nature.

Perhaps we need only to take the time to be comfortable within our own skin; to know ourselves. Perhaps in our searching, our world is much larger than ourselves and what we consider our family and community. This could be a guide for new ways of considering our world. But we need to make that choice. Just like with the negative, the same holds true for the positive, healthy, and supportive. Are we calling fear or are we mindful, remaining present and more balanced?

Life in Nature is sharing, adapting, understanding, and learning endlessly within and across species—all without a common word. These varying strengths and weaknesses create a more complete picture—a picture of our whole Earth, honored, including life, waterways, and communities throughout the natural world—in ways that are not shallow, but require great depth of intellect and are overflowing with information. There is more that connects us to the rest of life on Earth.

Held within our cells and in a very physical way, there is a mutual intelligence, and it is part of all of us. Positive connections, interactions, and relations strengthen and grow—just like the negative, toxic energy that can bring people down. But the positive is less weighted, and it happens faster, goes farther and deeper—if we remain open. Could we begin to accept and support the same in Nature? Can we begin to see that we have more in common with Nature than not?

As much as we would like to assign our connection to Nature as a connection to a specific, predetermined idea of what we need to be with Nature, there is no tangible location assigned to this reality. The area of consequence here is within our motive, our intent, directed to life in Nature—the fertile ground for which our exchanges with other people sprout. This reality resides in our hearts and minds. The power behind these influences will make or break our health. Our fluid, calmer, *trusted* mental states are the stuff of life in Nature.

Without the ability to create a calm, balanced dignity in my own mind and in my heart—the place of calm for which I am as I stand—I could not know, intimately, the flow of information through my awareness that comes from Nature. I used to feel very removed. The unpromising, unhopeful, and sadness from other situations were drowning me, and I did not know where I started and where another ended.

I was unaware of breathing through difficult times, how to release control that may or may not have been projected onto me; and honoring the right—my right—to have boundaries and feelings, and offering these same rights to all life around me. Feeling my way blindly, I did not offer support to myself, and this turned toxic. I was in a chronic state of ignoring reality, and I lived from this field of vision—even with good intentions—still trying to understand life alone, within. This lasted well into my adulthood.

We cannot be in that unruffled, present place that offers a slightly open door to a world ready to reciprocate—a place we might want—when the environment is unavailable. The environment created within may not be responsive, regardless of all the good we want and intend. For myself, I held this close, quietly. No one else and nothing else can create this environment. It starts and ends with each one of us.

Nature runs through our cells, our bodies, our minds, our blood—more than a lot of us care to admit. But we can change. It is our life to do so. Alike to breathing the air, it is possible to have it become second nature and become our Nature to honor Nature. Can we uphold our position in this equation? We will be forever changed.

We need to embrace our place among and beside these lives. These lives are too important in their own right. Our reality is ignored when we ignore reality. By not offering them the space for them to be, we are reducing our state of being, too. We cannot continue dismissing and ignoring the fact that Nature has a loud, dominant voice in our lives, our health, and our happiness. Together, we breathe from interconnectedness.

For our futures, something needs to change now, and the change lies in our hands, in our hearts, in our relationships—starting with who we are, as individuals, so that we may know another. This starts with rebuilding the relationship we have within the equation of all of life on our sacred Earth. If we want change, we need to be it ourselves first.

I thought that life in Nature was always in need of help. I thought they were helpless, and that I could do something to aid them from my idea of helplessness. I wanted to offer assistance. In moments I had as a child and as an adult interacting with Nature, I believed I was the one who knew what was right to do for these lives and what was needed. As if they were helpless children, I offered that help with all good intentions. But there was another side I had not considered. What if that individual in Nature was approaching me for reasons that I was unaware of? Perhaps that individual was offering his or her assistance, intelligence, and support. Maybe this life was letting me know what was needed, only to have it fall on deaf ears. What I did not understand was that the individual Animal or Insect, or the attention calling from a particular Tree or group of Trees, or a whole natural area in that moment, was not the one who needed the help and support. I did.

Searching, Still

Nature lives, understands, feels, and knows in ways that we have been striving for incessantly and have yet to obtain. Could Nature's role in our lives be much deeper than any other external experience? Could we begin to look to Nature instead of turning our backs? Do we care enough to know Nature on Nature's terms—perhaps in awe and admiration? Might we begin to see a kinship with these lives as we all exist here, on Earth?

In moments of struggle, we can stop questioning ourselves and seek the support of Nature, in kinship, for we are not independent.

Within each one of us, we hold the difference between what we are presently and what we will be going forward. And yet we struggle with one another in ways that go against anything that would create our own happiness. Could we be doing this to ourselves?

There are magical connections existing in our bodies' physiological roles and the arrangements performed by the brain and mental functions, and their effects on one another. The relay of information and the thoughts of who we are—these are not separate from our inner world. Our bodies do not always fail us. We are much more likely to fail our bodies. Just as we are one with life in Nature, we are of our bodies and what they can turn to—and return from—stemming from our minds.

Even as a child, the support I was looking for was within me, even if it was without me. Unsound motive and intent are not who we are in a healthy state. Unsound and unrestricted motive and intent decay our health, starting within our cells, and are destructive to us and those around us — including life in Nature.

Nature experiences equivalent information through similar senses. Nature knows, too. I believe that a heavy heart without much hope—feeling lost and without support—is a very real thing. Through our senses, these feelings run deep—deep down to that cellular level.

It is easy to forget the role that Nature plays in our lives. Like drinking water and breathing air, we do not keep this in the forefront of who we are to allow for the bond to blossom. But, like eating—and all life's universal need for clean water—we can learn to make it a priority and drink in more.

We need to breathe in the world around us to support our lives beyond our initial thoughts of what life in Nature provides for us—beyond food, materials, and medicines. We must *feel* with and for Nature.

Before I leave the house for a walk, I focus on our indoor air. I've tried to sense with it. It is warm and comfortable, but it lacks life. Outside, in the cool evening hours, the air is full of life. I've tried to ignore or "pause" my experience when I smell auto gasoline or food smells from someone's dinner preparation. When it is clean air sinking from the Rockies, it is colorful—even though it has no color that I am able to see visibly. There is energy and life; clean, free—the air itself is free. Messages from the mountains come through the air. Messages that I can't interpret—currently. There is a difference between free air and closeted air. Even air gets stagnant and cannot evolve in a closet. I sense other areas—places — in the breeze; places that have life carried from it through the air to where I stand. A world of information.

With all of our struggling—with all the struggling defaulting to life throughout Nature, too—the question now is: why?

We need to be living with our yards, Trees within our communities, life around us in places near where we rest our heads—in our personal equation.

Even if we have adopted this belief that we are superior—more advanced—is it not, then, our responsibility to care for these lives with even more concern and consideration?

Do we care too much about what other people think, creating the fertile ground to operate in a way we know in our hearts is constraining us and is not consenting to our Earth-bound connection to all life?

We receive information all of the time, coming from the bad and the good. We see it, hear it, are told it, read about it, feel it, sense it—and some may even know it. I've known it when I *felt* it. Given the value placed on other ways of knowing people and situations, I've done my very best to keep these physical reactions to people and situations to myself completely. And eventually, to my detriment, I learned to ignore my knowing. This can turn sour quickly. However, I still knew that what I felt was more reliable than any words expressed or actions that seemed out of alignment in a situation. Without knowing the terms, from early in my life, filtering information along the way silently, I learned how strong motive and intent are in relationships. My inner knowing filters these with great precision—and also with a gentle understanding. Still, these are, at times, not for me to speak. In this way, our internal knowledge and grounding are our strength. Who we are, and our reasons for our actions, are known by others through the air. We never need to explain this away to anyone, starting with ourselves. For some, we cannot be trusted by anyone, especially life throughout Nature, if we are not worth that trust.

We have messed up a lot. But that is the beauty. If we mess up, we can change using what we have learned. That is strength. The ability to be alone and be comfortable with our own thoughts and space is a daily necessity.

Regardless of what we think we might need, restless—just out our door—life in Nature turns us gently to focus on another, so that we may learn what it is we need for our calm. If our physical health is multifaceted and yet seamless through the mind—with life in Nature—then it is rational to consider multiple sciences and disciplines focusing together toward challenges we and our societies have as a whole. Including life in Nature at the table of our thinking and decisions will only help us—and perhaps be the only thing that will help us to the degree we seek. Not as a use for us to do with as we please, but as a voice in their own right, echoing within each one of us in response.

With a mind that is restricted, held to focus, and focused on life within and around us, equally—and with a mind that is intelligent enough to embrace the need for calm long enough so that accepting our physical, mental, and emotional willingness to remain open to Nature's voice and what is truly possible—our restlessness in a reality ignored begins to subside. Our struggles are softened, and a new but ever-present intelligence is felt in release.

The reality is this: these lives in Nature, in their own right, are far more important than most of us have ever considered. The reality is, we cannot afford to ignore that these lives are central to each one of us in ways we have never realized. At the most apparent, our lack of attention to our personal, kind, and considerate connections with the lives that are Nature — and entire natural communities—has knocked us all down quite a bit, and our struggles are now becoming everyone's struggle.

We are each responsible. We each radiate out to the world. We each can create a world of change rippling out from us—on our porches, in our yards, on our streets, throughout our communities. That is the way of interconnectedness.

As a child, this is what I wanted. I wanted to feel kinship—kindness and love with Squirrel on a deep, real level. I yearned for a trusted relationship with this member of life in Nature. I trusted *that* more than anything I was told, witnessed, and was placed upon me without my permission. With Squirrel, we would understand one another. We would be understood.

Said within the quiet of self, but also with certain conviction: "Someday."

Two

Remembering the Collective Need for Boundaries

The Bridge to Nature

There is a void within us, and it is reflected in our behavior.

It has been revealing itself for some time now—in how we treat each other, how we treat ourselves, and how we think about who we are.

We are entering a time when holding life in Nature as part of our personal landscape is becoming not only acceptable, but necessary. This is not a new need. It is one we have always carried within us, but one we have largely forgotten.

Without personally welcoming Nature into our lives, we fail to recognize that these are individuals—other beings—trying to live their lives and meet their needs, just as we are.

Perhaps this is what has been missing all along.

In the midst of our current human struggles, and given the fragile state of so many wildlife populations, natural places, and wild areas, the time has come *now* to embrace life in Nature in a way that taps into our emotional strength. Perhaps we were designed to include these lives as part of our own inner environment. And perhaps, if more of us did, we would care more deeply about Earth—about the nonhuman lives who share this planet and need to live here alongside us.

Maybe we would care more. Perhaps we would pay closer attention than we do now, and act with greater compassion and restraint. Maybe we will begin to recognize that we are not separate from this living equation, but woven into it. And by neglecting it, we diminish not only the natural world, but ourselves.

In my twenties, I thought a lot about the barriers placed in front of environmental conservation efforts. I tried to understand or know the way to reach people more effectively. Why was it so hard when the connection felt so real? I wanted to know the root cause of any issues—what was preventing so much good? So I did what I do best: I tried to *feel* it. I read, listened, felt, learned, and wanted to understand more. I thought about the individual, about what inspires one to act in ways that include Nature. I wondered about what was holding back the change needed. Inspiration, I sensed, is still key—for us as individuals, too.

The divide between personal care and care for Nature continued to saturate my daydreams. It made me uncomfortable. There is a bridge that connects people with Nature—a bridge that seemed so clear to me, yet one I struggled to put into words. I could certainly feel it. So, what is this bridge?

All meaningful relationships that are healthy and good for us include considerations that are just as relevant to our relationship with Nature. I began to wonder how these same considerations exist among Nature, too. These influences are personal. I always knew, deep down, that we need Nature much more than we realize. Our mental health—my mental health—and our happiness are linked to Nature. But I wasn't sure how. At the time that I was first getting to know *the bridge*, I had no idea how an honest connection to Nature was vital to our purpose and our greatness.

I wondered—if we don't care about Nature, or if we don't value these lives and communities, then where do we draw the line in our care for our own lives? What is it that we value?

I focused on trying to understand this. I spent many quiet moments over the years feeling through these questions. Mindfulness helped. The years passed, and still I was exploring these deeper considerations.

Feeling I was in two different realities at once, I attempted to understand what was happening around me. I wanted to see through the outside anger and uncertainty, the inability to share openly, the attacks born from insecurity, the little value placed on trust—because none of it sat right with me. Almost everything felt uncomfortable. I read a little, observed a lot, and tuned in to a knowing that came through images, sensations, and feelings. I shared all of this information with no one. All of us have lived through times in our lives when everything around us did not match what we felt within us. For me, I felt there was something—or someone—supporting me, with me, through this.

Perhaps humanity wouldn't be in this current situation if we had remained intertwined with Nature—or evolved more deeply with life around us. We ignore the vacancies within ourselves that only life in Nature can align. Others are beginning to feel this shift, too. Change clears us out of our own way—together.

Some of us are taught that we are not worthy of boundaries. That holding them dear makes us wrong. Even if the words are never spoken, we sense this. Others act as such. But nothing is farther from the truth—or less natural.

We're out of balance. We disregard others. Many of us struggle to accept even the most basic necessities of boundary and space—in ourselves, in our relationships, and in how we treat the lives among Nature. We knowingly cross physical boundaries, seeking attention through orchestrated images we feel obligated to share. We ignore Nature's boundaries, so why would we consider our own—or those of someone we've never met?

Without boundaries, some may often dismiss them in others—perhaps unknowingly. And in doing so, we open ourselves to the hurt, fear, and insecurities of others pressing inward. When some cannot uphold our own boundaries, we can confuse our needs with another's. A disregard for physical, emotional, and mental space leaves all life—human and nonhuman—feeling overwhelmed. Overall, there is no need to create a distinction between people and life in Nature.

Do we want to feel drained of who we are, just to preserve how others might feel? Do we need to suffer for the wants of others, especially when those wants are not what we can—or should—give? Dismissing our boundaries places us in the same situation as Nature: neglected, used, taken for granted, and ignored. And then we return for more. Sometimes we turn on one another.

Our actions impact us physically. We cannot separate our thoughts and actions from our emotional and physical health. Our mindfulness deserves so much more love than we currently offer ourselves. Waiting, it hides in the corner. Nature watches.

Some of us feed Birds to provide food options during times needed. We enjoy their presence. If we allow for it, we learn from them. Some of us come to know individuals at our feeders. Bonds form. In moments, we get lost in their ways. Quietly, we admire them. We focus on their feathers and body shape, the differences between them in physical attributes, and marvel at their abilities. We stand in awe at their quick and observant experiences. They almost seem to know us. They test us to see if we will honor their boundaries, and perhaps the boundaries of their babies, too. Their actions, communication, interactions, and light pull us in on wingbeats and sound into another world. *Their* world. We want to know more, understand more, and learn more. We miss that, perhaps, they are reaching out to us, too—to include us, understand us, know us. Maybe they wish to share what they know. Maybe they are even teaching us about ourselves.

On an informational website about birdfeeder cleaning, one person referred to Birds in their yard as "my Birds." I do the same. They are part of my family. I care for them. I provide water and resting places—even in my very small yard. I do this because so much has been taken from them. I do this because their presence offers highly intelligent information that I need each day. But the need is deeper, still. I do it because they are worth their existence regardless of my needs.

I care, like many do.

One day, I was researching the proper birdfeeder location. Birds and I need to exist in my small yard, together, since we meet here daily on common ground. Their world is our world, too. A shared space.

Birdfeeders are very easy targets, especially in winter. While some Birds migrate because they eat Insects, solely, there are some who eat Insects and seeds, and those who eat only seeds. The energy in this nutritious food helps to keep Birds warm. In this moment, a few people were hoping to deter a local Hawk or one migrating through from using the feeder as an easy source of food. No one was looking to rid anyone, but to help create a balance.

Following some suggestions offered by others who create space for these lives, one commenter deflected the intention and took issue with this phrase, "my Birds." Unsolicited and out of context, this person stated that they were "tired of people claiming" Birds "as their own." "Hawks are just being Hawks," they explained—"this is the order of things." It read emotionless. "Let it be," it said. "Stop making it personal."

But although words mean something, the energy behind words involves more. In this case, I felt the energy behind the commenter's words as saying, "Stop creating relationships where there should be none." Another reader felt this way, as well. This commenter took offense. Then, the original person who used the phrase "my Birds" was not happy with the indifference, and from there, the exchange turned into something else entirely. Expletives were swapped. The conversation became very human and nothing of Bird life. And the written words ended just as abruptly, but the energy of the words remained.

If we can start to understand *the bridge,* we can understand the frustration that arises when Nature's boundaries are broken. We can also understand the need to be a support for those in need in Nature—because Nature's boundaries were *already* broken. We, humans, are the cause of this imbalance. We can be forgiven. We, too, can be seen as individuals among so many.

There is something to the way life in Nature embraces forgiveness. It seems that when boundaries must be broken for basic survival, the need to learn from the situation and continue living is vital. There are no words or exchanges of apologies in the ways people half-heartedly use and, at times, abuse. Forgiveness does not require the obligated response of "It's okay," or "Don't worry about it." In more tragic moments, forgiveness is present, and it sees the way to our future. But trust is something else entirely different. Life in Nature gently instructed me on this, too.

Boundaries and Belonging

We are designed to connect. We are also designed to survive. These instincts work in concert—until they don't. At some point, for those of us who are sensitive, it can feel like connection and survival are at odds. We yearn to belong, but at the cost of our own peace. We reach for others and then abandon ourselves.

But this is not true connection. This is self-betrayal, a survival strategy we were forced to learn in childhood.

True belonging requires boundaries. And the truth is, many of us didn't learn what boundaries even were—let alone how to embody them. What we learned instead was how to endure. We were taught that being loved and accepted meant staying quiet, keeping the peace, or tending to someone else's emotions at the expense of our own.

To survive, we've disowned the very sensitivities that once helped us feel connected to life. We've learned to be palatable, to stay small. Or to accommodate, even when it hurts. But the body never forgets. And neither does our sacred Earth.

Mason Bee

Mason Bees are pretty amazing beings. They look a lot like Honeybees but are solitary. They are pollinators. Regardless of the fear some people hold for them, they do not sting.

I have known one individual Mason Bee who frequents our porch near our two wooden chairs near our Lavender. This female liked this little area very much.

My husband and I were sitting in this space—in these two chairs—one afternoon. We were enjoying the sunshine. He decided to move one of the chairs. I told him Mason Bee was using one of the empty holes in this chair where there was once a screw. She was building layers of her den and food, with her eggs layered in, so her little ones could grow and evolve to emerge when ready and live their own lives among our flowering Plants. He assumed that by moving the chair, she would not mind so much.

As we continued to talk, Mason Bee was flying around in a frenzy. She flew up around his eyes, back and forth, and around his head. I knew she was trying to communicate with him. I have had Bees look into my eyes at times, too. She was persistent and determined. This went on for many minutes. I was watching her attempts to communicate. Not wanting to tell him what to do, I continued observing. He was not listening in a way that she needed. He was focused on his words and world. I soon paused from listening to him and focused solely on her.

It could have been timing, or perhaps she sensed my focus—life in Nature seems to know our focus and interest—because she finally gave up on him and came over to me. She bumped me twice with her body in the way Bees do to warn one another. I am familiar with this, but I have never experienced it personally until then. Without thought, I sensed a need from her. Again, she could not sting, but she was asking for help. I knew this instantly and said, "She wants you to put the chair back."

Having witnessed her interaction with me, my husband returned the chair and pulled up a different one. Instantly, she stopped panicking and flew back around the corner of the house.

I stood up and started to say out loud, "She'll know I am moving her chair again." And there she was—straight from around the corner up near my head and face, back and forth. I was only shifting the chair's angle oh so slightly. Once I let go, she buzzed around it, and when everything was in the proper place and her boundaries were respected, she went back around the corner of the house to continue her work gathering what she needed.

Her reaction made us both laugh. We sat and discussed her magnificence with wonderful humor at our own inability to see how our actions were disrupting her world greatly. We researched how Mason Bees live and what they do in their days. She was an amazing individual. And she would fly back every few minutes to go into her hole as she continued taking care of her needs—the needs of those she was preparing for—and was very content that we also allowed space for her to belong.

The Body Remembers

For a long time, I didn't understand what was happening to me. I trusted myself and what I knew and felt. Yet, I was told that I was too sensitive and, at times, it felt like I was. So I blamed myself for not being able to "handle" things the way others seemed to, even if that assumption proved wrong over time. I blamed my mind. I blamed who I was, as an individual. I tried to change because I was told to change. More than once. Sometimes, I projected outward, placing discontent on other people. I didn't understand that I was taking what was dumped on me and trying to rid myself of it. I tried to "fix" myself. Eventually, I really was unable to understand how someone else could love me.

But no amount of trying to "fix" the parts of me that others did not like, or adjusting myself to become someone who might make others happier—as I was often expected to do— would ever create a better version of who I am. Instead, it only caused me to like myself even less. In time, I came to realize that I do not have that kind of power. *I am not responsible for making other people happy.* Instead, I began to listen.

I began to follow the thread of what was actually happening underneath all the shame. I listened closely, in silence. *There are no words in the quiet.* It was so quiet that I opened doors to hear a knowing when there was something to know. It was so still at times... I could feel the understanding. I relied on this more than anything.

What I discovered was that I wasn't broken. I was absorbing. I was feeling too much because I had no safety for my sensitivity. I was untethered. *Unprotected.* And what I needed— what I'd always needed—were boundaries.

Boundaries are not the rigid, defensive kind that are more like walls and roadblocks. Nor the "I'm fine, leave me alone" kind. I needed boundaries that honored my body, that honored who I am and how I experience life—ones that honored the rhythm of my nervous system... that which honored the way I am designed.

I needed boundaries like Tree needs bark. Not a barrier, but an edge between myself and the world that I could control; something that can breathe when safe moments allow me to be who I am. Boundaries are something that provide protection for our personal strength. They are a way that protects what is sacred and allows life to move freely with ease. They are something of us and beyond us that guide and ground.

I was told I was too sensitive. At times, it may have seemed like it. It's confusing when we are young, and people's words do not match what they are really feeling. And I rarely relied on words to know feelings. Even on happy occasions that are full of emotion, often in a full room, I feel it all. I understand this now. But in the past, once I could no longer hold back, the dam would flood, and I would feel it all. Tears came easily in moments of great joy, in sadness, in loss, and in memories. The tears were mine, but also for everyone else feeling so strongly, and they were all filtering through me.

True to form, my ability to feel would also open the door to others projecting their own issues onto me. In addition, they would tell me I was wrong or that I was embarrassing them. Or worst of all, that I only wanted the attention. In truth, all I wanted was to be left alone in peace.

There were typically no questions about whether or not I was okay. There was no desire to understand what or why I was feeling so much. Instead, I was told who I was and what was wrong with me.

My body remembers moments very well. Now, my need to be understood is fulfilled instead by self-understanding—what I needed all along. And those difficult moments have provided so much light on so much truth. For that, I am grateful.

I am not alone. *We* are not alone. Our body houses who we are—always.

Today, self-trust and boundaries are my greatest strengths. And I can comfortably sit with myself, in my quiet, present in my body, and be at peace. There is no more need to try to explain myself away when I feel—my body and mind ground me, and I am safe.

This is what Nature has taught me and continues to teach. Boundaries are not only valid with life throughout the natural world—they are necessary.

Nature doesn't build walls. These lives and communities build shields and sheaths. They create layers. They uphold ecosystems that are interdependent, supportive, and intact.

And we are no different. We are Nature, too.

The Gentle Strength of Reclamation

We can hear and see quite clearly, but more often, we do not know how to *listen*. Perhaps there wasn't anyone in our lives who could attempt to model how to explain why. Different can be lonely, as I was once told. So, we continue to learn to override. Or to cope...or manage.

But coping is not the same as healing. And managing is not the same as belonging.

It can be difficult to stay in the present when our footing is held back. But when we stop struggling long enough to calm our minds, we can feel the flow within the currents of the natural world through us…if we respect their boundaries just as we need our own.

When we start to reclaim our boundaries, we begin to notice all the places we once abandoned ourselves in the name of connection. We start to feel the places where we gave too much, held too much, or stayed too long. We begin to feel the grief of our own self-abandonment.

And we begin to understand that our sensitivity was never the problem.

Our sensitivity is the messenger. It is the part of us that says, "This is too much." It is the aspect that whispers, "Something isn't right." It is the piece of us that wants to *slow down*, to rest, to stay close to that which is real. It's the part that knows what to do and where to go in our lives.

So, belonging begins when we are home within ourselves; when we start to notice what we feel, what we need, and what we desire. When we stop abandoning our inner landscape for the comfort of being accepted by those who hurt, and when we stop abandoning our bodies and our sense of self in order to be loved—this isn't easy, but neither is staying stuck.

Learning to listen again—to our bodies, to the breeze and hum of Earth, and to the deep intelligence of our innate sensitivity—this is how we begin to remember. It's how we come back into rhythm with life. Nature waits.

Where Grace Meets Dignity

When we begin to set boundaries, we feel the boundaries of others. Boundaries bring clarity. They bring us back into relationship with ourselves. But they can also bring discomfort, especially when we've spent our lives ignoring our own needs. The discomfort of doubting our right to our boundaries is soon replaced by respectful relationships, including the one with ourselves.

The natural world, which thrives among boundaries, shows us how we can learn to feel empowered by them. In time, we may understand that not having boundaries is unnatural.

At first, it may feel uncomfortable to say "no." The first few times might even feel abnormal. It may feel harsh or wrong to speak up. It may even feel selfish to protect our peace. But this is only because we were taught that our needs were "too much." That our feelings were "too big." That our boundaries were a burden. That our worth was not with boundaries.

This is not true. Stated again: being without boundaries is unnatural and not of Nature.

Our needs are not too much. Our feelings are not too big. In fact, feeling life and all the emotions is our strength. Our boundaries are not a burden. They are a gift. They are the gateway to connection. Boundaries are the soil in which safety grows, for us and for others. They are the structure or shield that allows our nervous system to soften. Here, we listen cleanly, openly, and with beautiful, healthy emotion. The stir within begins.

Understanding the bridge to Nature begins with boundaries. This is not about fixing ourselves or others. It's about remembering ourselves and that we are a part of something greater. We are allowed to take up space—we are meant to be here on this beautiful Earth. We are allowed to rest. We are able and expected to say "no" when needed and necessary. We do not need to explain ourselves away to anyone. We are allowed to need and to have needs.

When we honor our boundaries, we understand more clearly how to make that bridge to Earth, to all of Nature. Because our body is of Earth. And every time we choose to stay within our sense of self—to listen, tend, and protect—we are practicing the ecology of belonging.

We are re-weaving our place into the web of life. We are not independent from life in the natural world. Nature waits, silently.

Three
When Nature Flips the Mirror

Consciousness Reflected

There are days on the weekends when I would like to sleep longer than my usual early rising. I have always loved mornings—that quiet time outside remains my favorite part of the day. Sometimes I do stay in bed longer, even though I feel worse when I do; I wake up foggy and groggy for longer. But I prefer to feel well within myself, so I get up early because that is what I need.

During seasons when Birds especially benefit from extra support, I also rise for them. I take clean feeders out of evening storage—even if it is only a small feeder—and in the dark winter mornings when they most need added nutrients for warmth, I set up their feeding area with food and fresh water.

They still bathe in winter. Clean feathers allow them to remain fluffy and stay warm. They drink the water and appreciate my actions and assistance, even in these small ways.

In the spring, I do the same in the beautiful morning light. During this season's surrounding songs, I know Birds are planning for their babies and need more nutrients. Again, I do what I can, even if it is small. I know they appreciate my actions.

When their babies arrive, I learn how Birds care for them, feed them, and teach them the importance of bathing. Even though their little ones are very reluctant to jump in the water, the adults lead by example, and the young eventually follow.

All these actions are mine and by choice: I make these small contributions to life around me for them because these moments help greatly. This matters to me because everyone needs assistance at times. These actions also save lives for those who are older and cannot find enough food as readily and as quickly as their young.

Over the years, I have learned how my body responds to what I do with others in their presence. Senses are always at work, even if many have been dulled. For me, especially with life in the natural world around my home, I can feel my senses burst open and search for connection with others in the quiet morning air. The emotions, thoughts, and feelings from these senses are felt within me as a soothing flow of warm honey drifting through me slowly.

I do small actions for Nature because I care deeply for them. Within me—my body, cells, nervous system, and thoughts—my inner Nature responds to what I am doing, and I feel the effects physically, mentally, emotionally, and consciously. I am more alert, calm, present, and at peace in this sense of interconnectedness, even through the smallest contributions. My brain responds to my actions in the world, and my body responds to what my brain processes. I can feel it, and it feels like a very deep love.

Our actions trigger physiological responses within us and send out energy into the world. They have an effect and an impact. They are witnessed by others and sensed on that same energy level. They harbor our motive and intent.

Life Is Worthy of a Positive Reflection

The impacts of our actions implicate our physical selves. Our minds, our consciousness, our cells, and our corresponding physical states naturally thrive on the positive; our positive interactions and expressions serve us very well—and remaining mindfully aware never goes unnoticed in the natural world. In a favorable reflection, our bodies feel at home again. As a complete community, we will all be willing to extend true trust to one another.

Yet our thoughts continue to resist. Perhaps this is because our struggles are reflected in how we treat those we forget to value—or do not value at all. We all displace anger at times, some more than others, but the way we treat the life around us shines a bright light inward. Our relationships with others reveal our true feelings about ourselves for all the world to see.

When we think of another life in demeaning terms and in ways that place sole focus on our own individual desires or needs, we also treat life in Nature in adverse ways. But we can change our perspective and appreciate life in Nature differently. We can adopt a willing attitude that all life in Nature holds an inherent good and belongs to the larger community of life. By supporting one another, we honor each individual's rights and inherent value. With this perspective, the image we hold in our inner vision becomes more accurate and more inclusive. The balance we cultivate within can then radiate outward, no matter where the mirror is turned.

If we react negatively to the idea of respect, perhaps we need to look within and ask why. If we are too overwhelmed to treat these lives around us respectfully—in our homes, our yards, and throughout our neighborhoods—where does that leave us? How and what can we do to feel more peace?

Nature is an excellent teacher. If we are too tired, how might this be a factor? If we find it all silly and insignificant, then we are doing the same to ourselves.

Empowering our lives and communities with Nature does something to our internal landscape and outward considerations. How might this affect our unique and deeply personal mental and physical health? How can our actions support a more balanced and realistic image in the reflection of Nature?

We can start by choosing those in our lives who support us, instead of deciding to be around those who control and inflict ill will or mistreatment. Nature watches people harm and hurt, pollute and destroy.

We can be so much more of ourselves when we walk *alongside* life in Nature, continuing to value boundaries and creating a world of Us. Our strengths, loves, inspirations, and sources of support rise to the surface when we allow ourselves to feel, through our senses, the presence of life in Nature. The balance that Nature seeks to create begins to spread within us as well. In mindful moments of pause, we may come to see that we are never truly alone. We find support across every dimension of our well-being.

Life in Nature seeks this same balance and is worthy of the same care and reflection. Can we offer them the same in return?

The Mind that Mirrors Nature

We could keep ourselves emotionally distant from experiencing and feeling a deeper connection to the lives found throughout our natural world. We could ask ourselves why we need to do this, too. What role does this serve? More importantly, *who* does this serve?

Without mindfulness, we place ourselves, our emotions, and our personal world at a distance from life in Nature. Yet, Nature functions best when the collective whole, including ourselves, is understood as our unified consciousness.

Because we express ourselves differently does not mean one is better than the other. We all have boundaries, and there are differences in those boundaries and how they are expressed. Hold this close within. There are few who would welcome their privacy being impeded upon and disrespected. Many, for example, do not like cameras shoved in our faces or welcome others to stand close to us, or try to touch us, or worse, get very close to our children. Why would this be any different for those seeking peace? Why would this be any different for individual life throughout the natural world?

We can choose to be fully present and enjoy a moment through all that we sense and feel. The very fact that we feel compelled to capture an image suggests that the moment holds meaning for us. Yet in our effort to preserve it, we may miss the opportunity to form a deeper personal connection with the experience itself.

Images are flat and lifeless among so much life. Let's honor the boundaries of another's distance emotionally, physically, perhaps spiritually, including life in Nature. Be physically responsible and emotionally convincing, where distance in the right of boundaries actually brings us closer together.

Could boundaries and healthy spaces of all forms restore some of what we are missing within us? Let's continue to explore boundary respect in our daily ways, words, and special experiences, which creates fertile ground for healthy relationships. They hold the tone for our motive, which is also represented by our intent, and they allow for proper distance in our inner condition.

We are all unique individuals. Why not honor our needs as unique? Why not treat our homes and communities with value, where the lives who live and visit need it, too? Consider them. Treat these lives better. Through them, we could learn how we might treat ourselves better.

It is time to begin to let go of what *we* solely want and consider what *other lives* might *need*. Remain mindful that our actions and others' responses do impact our inner world, and that our attitudes impact everything.

If we can relearn to hear Nature's own specialized voice, the distance between one another within our emotional space, within our reactions and thoughts, and within our bodies and minds begins to fade. Perhaps this will create wholeness as an individual, together with all life on our Earth.

Missing from Within: Unfit for the Environment

So many of us understand the intrinsic value of Nature. However, many are unaware of the undercurrents: the unification, healthiness, and helpful information that individuals throughout the natural world create and utilize. It is as if we left the conversation too soon. Nature largely remains unheard.

We no longer have years left for change to take hold. We cannot assume someone else will balance life or the systems on our Earth. Collectively, by choice, we declare our instability to all other life around us. If Nature has shown us anything, it is that those who cannot adapt to uncertainty and hold unsustainable views will have this balance decided for them. Life will go on, and the equation will remain intact with or without us. The uncertainty in our personal views about ourselves and Nature, adding collectively to the human race, resonates throughout this equation.

We have generally lost sight of what it means to have deep, personal, kind connections. Individuality has stepped aside unless it comes from someone we deem to be untouchable; in which case, we tend to become followers. In essence, we have adapted to alienate one another. We could have great strength in diversity and intelligence, yet we are fed constant denigration and drawn into others' remote worlds on screens, losing awareness of our own inner conditions.

The environment we have created within ourselves is unfit for life in Nature. We have made inner spaces that exclude Nature, where the natural world is not welcome. For a long time, this has been taught to us as we've learned to focus our energies outside of ourselves to improve how we feel and to take what we need for happiness. We welcome quick sayings like "only we can make ourselves happy," yet scoff at what that might require.

We struggle to adapt to changes within our environments. Rarely do we address root causes. Instead, we add more vices and distractions. We only face the root cause when no quick fixes remain. Sometimes we just stay—stuck. To some, pain must be felt repeatedly before we realize our reactive responses. This pattern repeats in our lives, our health, and in Nature.

With boundaries intact, Nature is the support we have lacked. How we treat the natural world reflects how we treat others and ourselves. And as we have gradually turned away from this mirror, we turn our backs on Nature, including ourselves along the way. From this internal environment, we are without support because we have created this space.

This is about a lack of self-trust. Missing our own trust, we isolate ourselves from the life we need psychologically, emotionally, and physically. From our center, life struggles, and we approach everything off-balance. Nature remains weary.

It Begins from Within

Each one of us needs support. Sometimes we accept so little from others because we don't realize we need more. What we accept from others is what we believe about ourselves.

We can become immune to attempts from those who might serve well to guide us, but instead seek solace by working to break us. Preoccupied with making sense of it all, we can pull in close to what we value most. And even in these personal forms of protection, when the flood is too much, we drown in and become what others experience, unaware that negativity breeds only more negativity.

Looking up at the light shining through Trees, I never considered seeking those who brought me joy by just *being*. Perhaps experience distorted my understanding. Time with life in Nature became worth more than I understood.

As a child, I sought connections with Nature—like Squirrel on the phone line attached to the house. I wanted to know and understand this particular, individual Squirrel. Yet, when I first encountered Squirrel, I lacked what was needed within me to begin a lifelong conversation. There are two sides to this story, as with any story. First, I did not approach Nature on Nature's terms. That is because often, I felt alone, unworthy, and witnessed little stabilizing kindness, love, or respect. As such, I thought little of what I had to offer because I didn't realize I was more.

During moments that test us, we all need support, regardless of who we are or the form in which we exist. Conversations and relationships are never one-sided, no matter how much we try to suppress another's voice, including Nature's voice within us. Many of us have lived through moments that create what feels so unnatural within us. These same moments can create more stable ground as we go through life during those inevitable times when life shakes us at our core.

When we do not honor deeper connections that satisfy and stabilize our physical, mental, emotional, and mindful parts of self that complete who we are as individuals, thoughts, intentions, words, and actions reflect this in moments over time, and life suffers. Our own lives suffer in ways we can notice, too. We seem to have lost an overall belief in ourselves, relying on others to confirm who we are and what we need.

Over time, I realized a large part of me wanted to help life in Nature or people, especially children. In my mind, both collectives lacked a voice. At the time, I saw a distinction between these two worlds, but I wanted to shed light on their voices. But I neglected my own. I was aware of much, even if people found entertainment in my struggles. Without boundaries, I thought it was kinder not to stand up for myself positively, even if it meant removing myself. I lacked a positive voice. Without support for who I was, my hopes, dreams, health, and positivity declined within me. So, Nature had little voice within me, too.

I wanted to do great things, but I see now that the greatest action was balancing what was missing within me. Unaware of Nature's support, I closed doors within to blend in, to disappear. In those times, when I did not create inner space to support relationships with natural life beyond myself, I limited what I could give myself.

Sometimes, our voices are smothered along the way. Nature does have a voice, and we either accept or reject their support by allowing them to be heard. Unintentionally, I did not offer space to balance within Nature. Because of this, all else in my life was affected. Still, I followed the deep pull when it became too strong to ignore. Somewhere inside, despite misguided trust and desires, I heard something and listened as best as I could and continued to follow.

Without natural, adequate support, we doubt ourselves. This lack of trust leaks into our worlds. Our struggles to communicate, to understand, and to get along with others show this. Historically, too many have taken without asking, manipulating words and situations—still visible today in personal examples.

Self-trust starts to shudder when we do not honor commitments to ourselves or care for our needs and dreams. Self-trust has nothing to do with selfishness, regardless of what those attempting to break through our boundaries might say. Selfishness is acting regardless of harm to others. Selfishness is using others or situations to fill any void, stunting growth in trust. When someone does this intentionally, there is a smell of manipulation. With windows sealed shut, the words and actions enter and begin to fill the room.

We are dominated by words, yet we often abuse, dismiss, manipulate, and ignore them, leaving words unreliable. This can leave us confused, alone, misunderstood, or unsure of who we are, why we are, and what we can be, depending on the situation. It can be hard to breathe with all of the words. We may become as unreliable as the words spoken and heard. Our nervous systems react to this anxiousness, and our physical selves may feel under attack, while our mental selves search for ways to support our emotional selves, trying to seek calm and return us to balance. Our mindfulness watches from the other side of the window, wanting only to find a way to pry the window open and let in the sunshine and fresh breeze.

Anxiety plagues many. The uneasiness of life stems from fears that can go unrecognized without little masks. Fear is the root of many personal and social issues and intolerances. Fear in our thoughts turns into anxiety vibrating through our bodies. Wanting a release of this uncomfortable energy, we turn on one another. Here, too, our relationship with Nature is profoundly missing. Here, where our own thoughts affect us and turn on us, life stays at a great distance. Toxic thoughts disrupt brain health like Flowers seeking nourishment from polluted water. The struggle is hard, and the wilt of beauty painful.

Negativity, anger, and harmful feelings are not easy on us physically. They erode us, creating unstable ground that further erodes. We feel poorly and often inflict it on others. This erosion is not natural—if it were, it wouldn't harm health, those around us, or create such toxic air felt in our presence.

Many of us are hurting with nowhere to direct pain but complacency toward disrespect for life within and around us. Without guidance, some equate dominance with confidence and control—ways that reek of hurt and insecurity.

So then, we create what we suffer from most—*a lack of trust*. We are missing what we need when in doubt. When untrustworthy, we are not trusted. Life in Nature cannot find home within us without fertile self-trust.

Our struggles are reflected in how we treat these lives who also need home, family, and freedom as we do. Our lack of trust in ourselves is mirrored in how we affect those around us.

If we cling to old, ineffective ways, refusing to change for the betterment of ourselves, relationships, and connection to Nature, we become unfit to support life in and around us. Earth's current state shows this clearly. We project negativity outward, depleting inner support. As far as we have come, we as a human species have become, in many moments, quite devastatingly toxic. It starts in the mirror. It begins from within.

The Control of Nature Blocks Evolution

Our world is very different from even only one hundred years ago. We have advanced in ways that do so much good. But we refuse to even consider what is beyond ourselves; that we are not the only intelligent and emotion-filled life on this innately unified Earth.

It's not for us to decide certain behaviors are "stupid," or that Animals, Plants, and Insects lack intelligence. It's not for us to decide for life in Nature.

We fight hard to keep Nature shut out. Childish. Needlessly necessary.

By establishing clear walls between ourselves and other life, we go against who we are. We struggle mentally, physically, and spiritually. We miss the beauty within and are blind to the beauty just beyond our door. In uncertainty, we turn our backs.

We have overcomplicated what is basic. Holding on to control of Nature blocks our evolution. Our rigidity prevents us from returning to center.

We refuse help, even if it comes from the very lives we see as inferior.

The problem may be ours. Those who dominate and force change lack lasting power. The quiet, gentle actions flowing in harmony with Nature's equation hold true strength. Here lies intelligence unhindered by fear, hate, and control.

Nature adapts readily and is mindful of who we are and how we relate. Imbalance ebbs and flows, but the need to connect remains. No matter how we live or what technology we create, there is no substitute for life in Nature within and through us.

We need these lives not for what they do, but because they are worth it. Until we alter how we view them, we remain distant, believing we are better.

Ignoring Signs from Nature

How we treat Nature mirrors much of what is going on within each of us. We ignore signs from Nature—within our bodies and the world around us—that could help. We seek "escapes," but what are we escaping? Ourselves? Our lives as we live them? The lives we have chosen and created? We do not generally take the time to listen and learn from Nature. The signs are all around us, and too many are struggling to stay afloat.

We undervalue, poison, and cause pain to the lives and living systems that may seem to exist at the periphery of our own. But when we replace negativity with a positive thought, words, feelings, phrases, or visions experienced with an individual life in the natural world, it does something powerful within our core.

Our Inner World Reflects Our Relationship with Nature

Many people going through their own struggles have shown me their true colors—friends, acquaintances, strangers—through kindness, harsh words, and cruel actions. The kindness is steady, strong, stable, and focused on the endpoint of something sound. The harshness is of fear in biting form. Before, I absorbed it all; now, it is not mine to hold.

One night, I sensed something, like a sneeze on the verge but not arriving. I stuck with the feeling because letting it go wouldn't ease it, and I disliked how it felt. Being out of balance sometimes is human, but not okay for me. I have learned to comfort, inspire, support, and encourage myself—this is how I understand, sensing through and alongside Nature, even when unaware. Now that I internalize these relationships, they grow stronger. When I don't like how I feel, I can't just accept it. I need to calm and be present with these lives—Nature needs it too. I don't need to be hard on myself for reacting to others' misdirected unhappiness. I need to return to the soft calm that makes sense—Nature. I must stay true to myself regardless of others.

I sat outside tending Flowers in the warm sunshine while Birds ate, then told my husband why I love who he is—a few heartfelt sentences, empowering to express, feel, receive, and release. We do this often. It's not about needing to hear it; we are both deeply grateful for our lives.

I am expressive when I choose, unafraid to be open, comfortable with emotion—even intense emotion. I have always sensed strong feelings from people, sometimes deafening, even when trying to ignore them. I offer kindness through words and sentiment. Some grow uncomfortable; others are moved to tears, as if hearing validation they may have lacked. I gauge and respond accordingly. I remain silent retuning to what worked well when I was a child. I have grown to take pride in my words. This was not always true. It is easy to drown in what we have always known.

We have to remember to surround ourselves with what we want in our lives. We become our environment, especially when unaware of the importance of protecting our sense of self. With Nature, when we offer kindness and peace, it flows back through us among the soft, warm breeze. Wisdom never requires a loud, dominant voice.

I'm careful not to assume what I sense in others matches their true feelings—just as with Nature. Our worlds differ. When I misassign feelings, I've been wrong. It's easy to confuse with our own reactions. I work hard to separate what I feel from sensed emotions. Anxiety can always arise from sensing others' anxiety, given the way it vibrates, even if I rarely experience anxiety on my own. Our boundaries become sound and strong when we breathe through the appreciation that everyone's boundaries are essential.

With us, with my husband, we can simply be together. When feelings come from within, expressing them is sure-footed and easy.

It's hard to make sense of the world when we deny our senses—community, joy, the light of our sun. We are not separate from life throughout Nature, no matter how much we argue otherwise. These relationships start from within—within each one of us, within our thoughts, and with the words we bring forward and release to the world.

Over the years since writing this, I've learned the honor in knowing not everything is meant to be shared. I go within, where knowing flows and is welcomed. Nature watches, aware of when we uphold self-respect of our personal truth and or place within interconnected intelligence. In this way, the good within each of us shines through, only desiring an echoing of more good—around with life outside our window, as well as within our brains.

Our brains process information through feedback loops, releasing chemicals that flow and flood our bodies. Our inner world shapes how we think of ourselves. The words used that our brains have created from our thoughts, or have allowed to slip from our mouths, have energy. These thoughts radiate from us outward to life—including Nature.

Even if unintentional, our missing-from-within state will not cease; physiology demands connection. Quickly pulled from centering, we fall back to people, places, situations, and substitutes we think complete us. Entertainment, hobbies, and so-called necessities are not our center. When missing within, we feel absence deeply. We search outside of ourselves in the wrong places, often void of life at all. Emotions from experiences meet on this shared bridge with life throughout the natural world.

Removing Nature from our inner landscape ignores much of who we are and our completeness. Taking care of ourselves goes hand in hand with self-trust. We lose our source of support and balance found only in Nature when we do not honor our part in this process—within the center, the foundation, of each of us. Little grows from an inner world that does not believe in itself. Nature operates in balance, not negativity. Self-trust cultivates wonderful positives within, opening doors around us.

At a cellular level, we are part of life's larger equation with Nature. Our physical, mental, emotional, intellectual, perhaps spiritual—and always our sensitivities—are our collective within, which ebb and flow among our experiences in unity with the natural world.

Most of us have limited connections that inspire and balance us daily. Without Nature, we will always be missing something. This absence surfaces in ways until we develop these relationships.

What we are within radiates out to life around us, with or without our awareness.

Nature's strength offers the support we need, but we must want it without conditions or anger. We must adapt to others with the same consideration we long for ourselves.

Life in Nature is brilliantly intelligent and bursting with emotion in ways many have never considered. Imagine what this means for our health and spirit—empowering camaraderie—if we welcome these always-present relationships. True welcoming means accepting that our personal experience is not the only way.

We, as members of this Earth, are just that—members of a larger community of life. We need connection with Nature, starting with the certainty that breathes within each of us. Perhaps it is time to awaken our connection within. We need the complete, total community. For now, we are missing a most basic need.

Stepping Through the Mirror

There are cells in the brain involved in processing information around us to help us learn and understand others, as well as situations beyond spoken words. These are called mirror neurons. I am not an expert here, but I am aware that it is believed that those who are more empathetic and, for some, able to understand without words spoken, have more mirror neurons. The silence speaks. It is further believed that those who are intuitively empathic have an even higher-level reaching knowing through senses. When in touch with their understanding of self, these individuals may rely on this information readily in situations and with people.

Some of us are more aligned with the natural world. Perhaps it takes others time to enliven parts of our internal intelligence to begin to break down walls. Those of us who are wired to do so have not lost touch with these areas of self from childhood.

I believe this *knowing* occurs in Nature—I believe this through my experiences with individuals—that understanding crosses species. Birds and Squirrels are more inclined to stop and know in return. Still, those who are more apprehensive have witnessed and experienced unwarranted human activity. As always, trust cannot exist with those who are untrustworthy.

Repeated attempts to engage with life in Nature unceremoniously can unintentionally disregard their boundaries, their physical and emotional well-being, and their need for space. When we allow them what they need—offering compassion, honoring their circumstances, and giving them the freedom to be who they are without requiring them to adapt to our wants or perceived needs—trust begins to emerge.

Perhaps to them, we are nothing more than destruction-makers and something to be feared. Perhaps it is here that we begin to see one another. Perhaps we begin to see what it is we need, want, and desire: to embrace who we are in a true state of simply being. Maybe then we will begin to trust, and the distance of space and time will dissolve.

We are more alike than we think. Squirrel, Salmon, Deer, Eagle, Squid, Tree—we are each uniquely different in our way of being, our presence, our experience, our contribution to the whole. We are needed—all life on Earth. Since we have failed to embrace interconnectedness, it is time to fall silent and take our seats as we let other individuals and communities express themselves. We need to learn through what we sense and feel. The more we are able to see the sameness within our uniqueness, the more we are likely to relate to and understand one another within our own right. Without interconnectedness held in high regard in our consciousness, our lives, and our actions, we suffer. From camaraderie, we trust in common ground, from common ground, and appreciation of each one's inherent worth—respect evolves within.

Cooper's Hawk in My Yard, Images in My Head

I can hear Blackbirds while I am in the main area of our home. I can also hear Birds if there is a lot of energy while eating.

I was reading something unrelated about an experience I had at a different home with Sparrow and Cooper's Hawk. This experience popped back into my mind... It was winter.

I knew immediately to go out back right away. I could not get there fast enough when I saw Cooper's Hawk corner Sparrow in the cage around our feeder, reaching in with her leg as Sparrow dropped down, mouth open, trying to protect himself. By the time I made it outside, she had him and tried to fly away, but he was struggling to free himself, and she dropped to the ground to get a better grip. She left the yard with him. It was all so fast. I removed the cage after that and have not used one since.

Fleeting images, memories, and pictures happen like this often throughout the day for me. But that day, there was a connection that I linked for the first time.

Not but a minute after reliving this experience within, I heard a warning call from Redwing in the wetland, one I have never heard before. It was so different and frantic, jarring, that I stood and went to the sliders to see what was going on.

In the late bloom of spring, there he was—red-eyed Cooper's Hawk, next to two small feeders in the yard—just as we were creating our backyard environment suitable for us to enjoy and suitable for them to live. He was on the wetland fence shaking his tail, looking. Quickly, I put on my shoes and went through the door.

He paid no attention to me. They often don't anymore. He was looking around and below for any movement to lock in on. I remained calm and stood there quietly. I tried to let him understand that the space is a safe zone for all to drink, rest, learn, play, and be. I tried to convey that he is welcome to rest and drink, too.

I have experienced this in other yards. He was new and young. He continued to look around the ground. All other Birds left the area, making warning calls far away from a group of beautiful, tall, and old Trees upstream. They were watching.

I decided to set boundaries and began walking toward him, slowly. He was about ten feet from me at this point, and as I walked, I was within six, and he flew out straight into the wetland—a large area with much food to offer him. When he left, a female Redwing and a female House Finch both shot off from right in front of me from hiding very still next to Plant I had placed there for cover in our very small and very empty yard, waiting for proper restoration. They left, calling in a frantic hurry off to the location of everyone else.

I watched Hawk fly over the creek and to the right into a little shrub area. I saw Hummingbird fly out and land in Tree a little away. Now, it felt as if they were letting him know his intentions were not welcome in the area. I have witnessed Hawks resting near feeding Birds without any disruption. The intentions were to fill an understandable hunger need, so it seems, among them. However, they never give up hope in life.

More Finches, Redwings, and Swallows who were still hiding, flew off to the safety of others. I stayed in the yard for a few minutes and then went back inside. After about five minutes inside, I went back out and heard a small chirp coming from just beyond our yard on the other side of the natural area fence. Someone replied farther off, and then five more Birds who were hiding in some Grass flew off, frantic and to safety. They were right below where Hawk perched on the fence. The Hawk had moved to a different area of our view.

Given our current world, we need to find ways to live together that work for all near our homes and within our communities. Crows and Hawks do this. Beyond my understanding, I have witnessed time and time again Crows working with Hawks to escort them to one side or another of a trail or a street, and then harmony balances again. I have had yards where Hawks come to rest peacefully while little Birds are active all around them, just as peaceful. I do not understand everything, but where I live, it is for all to be together. I try my best to learn from them to help our common areas we share, such as a yard. My motive is balance always, and my intentions are to create health, safety, and spaces to be and enjoy. I have had to be creative at times and learn to listen in ways without sound or speech. Always, the space seems to balance out.

After only being there for just over a month, we were visited by a lot of Birds. Back in the yard, right after these remaining five left from hiding and others met with them in the sky, Birds began returning to the yard. A lot. And they continued to come in waves.

The male Redwings were first, who keep watch in the awful heat, sometimes hail, and often rain. Female Redwings guard their babies endlessly and fiercely. Finches work nonstop all day long, making sure food is plentiful for their little ones. So many flew in, landing in the location where Hawk was on the wood fence, with some flying over the area in a circle about five to ten feet above, all chatting. They stop to stand on the fence for a moment, then they leave, heading back out the same way they flew in...and all doing it the same.

I stood there in awe among them, witnessing their actions. Due to the number of Birds and how they came back to the yard in such a special and unique way, with no one eating, I am certain something was being conveyed to me.

Afterward, no one came back to the yard for a while as I was checking from the windows. That was when I sensed gratitude out of nowhere. I do believe Birds flew back to me to express gratitude that Cooper's Hawk left safely; that none of their family were taken from a space that we all share carefully. They are family.

Information is delayed at times, once adrenaline returns to normal and thinking calms for the mind to know. Hawk is family. That yard is very small, yet surrounded by a larger natural area for Hawk. These are my boundaries when establishing a place for life in the natural world. These are my boundaries, working together with them without harm or breaking trust.

Since this experience, more Birds are coming. They are less nervous in the yard with me, especially when eating and drinking water. I am excited to have our very small yard finished in a way to support Birds, Bees, Insects, and Hummingbirds, so as to offer everyone more comfortable areas to rest and shade from our sun. There will be places to poke around. There will be more Flowers for Bees, Butterflies, Wasps, and Spiders. Now, even Hawk rests just beyond in Trees. Deer sleep in Grasses just outside the fence in the natural area; they leave large impressions that are not seen anywhere else along the other homes. Hummingbird sleeps through the night on our porch in one Plant hanging from above.

Perhaps they sense something from me, that they are safe and respected by me. I am mindful of them. They often play when I turn my back, and I hear them taking turns to get close to me, only to fly off when I turn and then do it again once I go back to weeding. They have a wonderful sense of humor. Perhaps the trust I extend consciously to them is known by them. They notice. Maybe it is all because I notice, too.

Reflected by Nature

Earth. Our Earth—for which all of the natural world is part, for which we are part—is beginning to warn us directly. What we are doing to life in Nature, we are doing to ourselves, ultimately. This is *not* us against them. It is not of sides for the taking. Our treatment of Earth, regardless of who and where, impacts Us—today. And each day thereafter. Today and tomorrow matter now, more than ever.

We *can* do something, and we can do it on our own, creating our together, going forward.

The wonderful thing about this is that we have within our power the vision and ability *for* change and *to* change.

We can create calming, positive, and supportive environments within our own worlds and places so close to our hearts, our homes, and our families. We can create and implement sound ways of being and reacting.

The treatment of these lives and our treatment of ourselves—our lives—hold strong and deep parallels. If we are the company we keep, there is no one closer to us than ourselves. What is that relationship like, and how might we improve it so that we want the same for life around us? It begins within and within each one of us.

Doing right by life on this Earth is a protection. Doing right is knowing that balance with all life is necessary, even if we never meet face-to-face. Clean air is something all of us need; clean water is something all of us need; clean and healthy soil, hearts, minds, bodies, and interconnectedness are something all of us need. The needs of all of Us are needed by all of Us—for each individual.

Dragonfly's Message

When my husband and I first began dating, we were visiting a property with a very pretty, quiet pond. The sky was a clear, bright blue, and the pond had growth around that area providing shade, cover, and life. It was beautiful in their calm quiet.

I asked him to focus on the pond. "Calm your thoughts, focus outside of yourself, and feel." Being one who is almost always in his head, he still tried. A few moments passed. He became moved and was in awe and understood everything I had been trying to use words to explain to him. He no longer needed the words to understand. The words did little justice, anyway. The experience was all sensed and felt, and with emotion.

I asked him to revisit that time for me, along with his permission to add it to this offering. He is more of words rather than feelings and emotions. This is a wonderful balance between us. Still, this is what he explained:

"What we talked about was intriguing to me because I was open to understanding more. It sounded interesting. But what I understood was the words. Still, the words were very linear, and that is how they are with people. It is more to convey what is needed and what will be done in a more functional way. More task-oriented."

He continued, "As I focused on the small pond of water, you told me, 'You are drawn to water.' And I am. A few more moments, and Dragonfly appeared. You told me to settle my focus on Dragonfly. I watched Dragonfly and was amazed at their abilities and color. You explained, 'That Dragonfly is for you. They have a message for you.' You told me to ask Dragonfly quietly and to myself with a calm heart, 'May I join you?' (A very crucial part of gaining consent when being with and among the natural world and a fundamental component of applied ecopsychology.) I did because I felt it was right to do so, and focused. I focused on them—the Dragonfly—to understand, and it was like a whole world opened." Using his hands while explaining, he created a hard wall of separation between both sets of his fingertips before touching and pulling apart an invisible line connecting him to this part of life in that moment. He then said, "The grandeur of the Grand Canyon is found in this little pond not far from home."

He finished his experience with, "The barrier between the pond area with Dragonfly and me melted into one."

If we have doubt or feel discouraged, let's face ourselves in the mirror of Nature to learn why. We have a world of brilliant people, with history and practices that have worked, and *do* work now. We are never too old, too young, or too "advanced" to evolve. We need these lives around us, and we need to work on repairing the harm we have injected into how we view and approach these relationships. We can help one another, Us together, as a mirrored and overlapping image.

Four
When Words Fail Us

All Emotions and Experiences Are Interconnected

Words fail us often. They fail me, too, more than I would like.

In my confident quiet, I find that people approach me when they are in need of support. They are in the depths of their emotions, sensing, and feeling. *Here* is my comfortable place. They share their fears and worries, and sometimes, their pain. They feel with me.

All emotions and experiences are connected. We all have a role. However, without a word, we are more like Nature than not. Beyond words, there is much more reliable information connecting us all.

Beyond Verbal Communication

When I feel something—I *know* about situations and individuals who may not be openly apparent. I was not (and still am not) willing to share most information. Therefore, my actions are often misinterpreted time and time again. In the past, this was partly my fault because I would never explain what, why, or how; I no longer saw the point in explaining, and I held on to this. I would not share, and I developed ways of coping to navigate through experiences when I was very young, in my teens, and up to my early forties. To this day, I still use some of these coping mechanisms, although I understand more of myself now. I became, or was willing to be, misunderstood to protect what I understood. I remain loyal to this with myself.

Sometimes, to protect ourselves and others, our personal experiences and paths are uniquely our own.

Previously, as someone who experienced much through other information *beyond verbal communication*, I shared none of it with anyone. I often struggled, closed and protected, to make sense of the flood of this information and these experiences. Through trial and error, I learned—and I remain learning—how to navigate these waters. Others would see the reactive side of me, only. I can see why it would be confusing. And I never defended myself; when I tried, the defense itself did not allow me to be as I am. Being protected, for one reason or another, enhanced my ability to understand, know, or sense other information. Because I focused on it solely, quietly, my understanding developed more and more. With this, so did understanding the wisdom in silence.

When we get comfortable with silence, it is very easy to understand deeper conversations and interactions.

People who are gifted in casual conversation are talented, in my mind. I can learn to fake it—and I did, for so many years, for roles in my career and in my personal life. But it feels like tapping on a wall to me, and I don't like how that feels. I have accepted that I do not do casual or "small" talk well, and I am comfortable in my quiet.

Insincerity is not the same as being uninterested. I consider myself more interested than "typical," given the flood of information that may, or may not, be ready for me—or meant for me. Remaining quiet, sensing, I turn to life outside my window.

Knowing Is Nature Communication

I assumed my support of Squirrel friend operated the same way. With people, I felt lost in a sea of struggles around me. I know what I knew, but I did not know it was a defined way of knowing until my late forties. With all of the work I did in therapy, following beliefs I developed by questioning everything and evolving what works for me, and growing in my acceptance that each of us needs to find what works for us as individuals, I always acknowledged information.

Recently, I feel most connected to sources. Pieces of my puzzle have finally taken shape to complete my picture. I am protected still and supported greatly. However, I have lived the majority of my life without a voice, without being clear with my voice, and leaving the wrong impression. I operate more unnaturally now because it is natural to do so—just like life throughout Nature.

Trust is earned. Just as information can be delayed, often, information is not meant to be shared.

People like me can struggle with words because words are limiting and rarely capture the essence of what we are trying to convey based on what we know and feel through a sense or our senses. *Words will never be able to capture how we know what we know.*

There are moments when we attempt to translate the knowledge of this *knowing* that is intended for sharing into words. We must honor the knowing of when not to do so.

I go through periods sharing nothing. Nothing at all. It is difficult to live in a world when it feels so natural to be very open, in kindness and care. Although they are not always apparent in moments when unsolicited, words matter. At the same time, they do not. But I understand why now, and that is helpful.

Those forms of *knowing through what we sense* are part of Nature communication, information, and intelligence across the landscapes of beautiful Earth. We know so little of this knowledge and readily struggle within this open-air arena. However, we are not removed. Nature is always trying to communicate with us in this way. It is up to us to stop and listen, without human words, and experience the other ways of knowing independent of our personal emotions and biases so that, perhaps, we begin to develop trust and understanding. It can be done with our human counterparts, but the need for this is rooted with life in Nature.

Perhaps it has never been my complete responsibility. I am guessing not, since there are always two sides to a story. I see now that what I was sensing was another's great success or emotionally stunted way of communicating. Even the twisted, manipulative, and deceitful words will always come through without clear light. I struggle with words because often the words coming through are of struggle. The struggle sensed and understood, when not of light, when twisted in hurt and fear, vibrates with anxiety in moments that cloud clear processes, and radiates onto others. We are responsible for ourselves. Without words, our ability to respond can become our strength.

Nature Is Completeness

For so many of us, we struggle with communication that is effective. Courses are offered, therapy focuses on it, so many of us are terrified of it, and a lot of us avoid it. Interestingly, verbal language is our main mode of communication—when spoken, we can be terrible at it. Completeness within us comes from knowing and accepting all aspects of who we are—the imperfections as well as our strengths and gifts. There is no shame in our limitations, for we all have special strengths. We can be moved deeply without words spoken. Are the lives that are collectively referred to as "Nature" so different?

Willow and I were walking on a trail we loved. I felt a distraction, but could not intellectually understand why. And I was unable to assign any words. At this point, I decided to head back home. Something appeared to be trying to attract my ear; my hearing; something wanted me to listen. I see now that my understanding was clouded by focusing on something instead of *someone.*

As we walked, it became sunnier and warmer. My phone said it was 74 degrees. And I had deep chills in me. I didn't know why; it wasn't like I was ill. But the hair on my arms was standing on end, even in the warmth of the sun. The felt chill lasted a good three minutes or so.

We continued to walk, and I came upon Tree uprooted by the wind a few days prior. Tree was still alive and wanting to hold on—a very big, beautiful Tree. I offered quiet reflection for Tree, like I do for fallen Animals. There is a peace with this action—in acknowledgment. For them.

This individual Tree's death was very different from that of any Animal getting hit by a car, tragically. One is very natural; the other is very human-inflicted and very unnatural. There is a big difference. There is no peace in Animal's death by car or Tree's death by human hand. This is important. Not all tragic loss is intended, and not all loss is in dishonor. Motive and intent step forward here.

As we walked back home, my sense of hearing was on overload, and everything was just way too loud for me. I continued to go home at this point to write.

And I wrote, "There is an old wisdom that we are disconnected, yet once reconnected... What's the word...*ugh,* having trouble putting the sensations into words. It is the same strength that I can physically feel when I put my hand on Tree, and I feel it: the strength and wisdom.

I am having trouble putting words to this."

And I wrote more later, "I'm back—a few days later—the sensation is: *completeness*. The word I was looking for is *completeness*—true completeness. Natural areas operate in completeness. They are *complete*. Perhaps not the best word, but it will work for right now."

I continued, "We can alter an injury created with incomplete motives; we can offer consideration for the individual. Natural systems only know that the way to be is *completeness*. Are we so different? And some suffer greatly because anything other than that isn't natural."

Completeness is part of Earth's natural process among individuals, groups, communities— all of life on our Earth. Remove an aspect, and it is felt in a ripple that grows amid our world, together. What creates this completeness is as natural as spring water and clean air, and we need it just as much—our basic need for healthy relationships with the natural world—flowing to us and through us, back to Earth. The senses of our interconnected world open another level of knowing.

Our Truest Support Exists Among Nature

We often do not interact well with others through our main route of understanding—verbal language. And when we are good at it, it is considered a skill. We believe we are set apart from other lives on Earth with an ability that we do not do all that well, generally speaking. Without embracing other forms of information and communication, we restrict ourselves through our limited relationships with Nature.

Our languages often limit us because of their limited nature. We can find ourselves searching for the right words—or any words—to express a thought, experience, emotion, belief, connection, or something else. There are moments when no words can express profoundness and times when words might not even exist to express or explain an experience. It is not because any of us are unintelligent or incapable. Perhaps we fear looking foolish when we seem to talk in pictures, or speak among analogies from impressions that create scenes, images, or described sensations. Words often do not do justice. There is more to our inner Nature, throughout all of Nature that surrounds and courses through us. Nature senses.

Through varying ways and the strength of sensing, life throughout Nature communicates well beyond their focused verbal calls and sounds. They communicate and understand through knowing, intuition, images, and all of their vast senses, even tapping into light waves and varying forms of energy—so much energy that carries and conveys messages and information. I will not even pretend to be a keeper of this information. It seems to be known among them, and I have only observed a few grains of sand in this desert. Perhaps we are not meant to know everything, except that all of this information is highly sensitive and highly intelligent. We do share some of these senses, and they are nothing of words and all of experience, feeling, and emotion.

Often unbeknownst to us, we do peek in and focus on Nature's other areas of communication (mostly verbal), such as when we feel something is not right and can't really point to why. Even though we might be unaware of the answer, others are picking up on our mood changes, true intent and motive, hidden sadness and struggles, and from our feet through our head, joy. How we are feeling or what we are thinking gives off incredible information to our surroundings without a word ever uttered. We, as natural beings, mirror this with life in Nature.

It is because of these other ways of experiencing life that we can have a truer understanding of another, including individual life in the natural world.

Perhaps limitations stemming from our own laboring skills in communication prevent us from seeing or appreciating individual uniqueness, so we cast our eyes on life as objects or things. Maybe we have been viewed in an unfavorable light regarding our needs. Maybe we were misinterpreted based on limited abilities to express ourselves in ways where we *are* heard. Perhaps, then, it is easier for some of us to sympathize and even understand the struggle between us and nonhuman life. Are we as different as we think, or is this something we want to believe because, for us, we find it very hard to see a way out of our own struggle?

I understand how tempting and easy it is to simply remain positive and think that a more sense-based form of communication is all that is needed. I understand how hard it can be to feel, especially with emotions that are not all that fun to experience—certainly not those we might be reliving from facing an issue head-on. It may have been awful enough when we were living through it in real time.

But we are never alone in our experiences. Our current situations call for personal compassion and acceptance. Life in Nature knows this, too. Looking away with positive thinking alone will not address what is standing right in front of us, blocking our way. We still feel something—no words necessary. Nature is waiting.

Life in Nature might offer us the support we never believed we needed, helping us out of our struggle—no words required. It is in our connection with an individual member of the natural world—on their terms, honoring their boundaries, support radiating from our deepest intent, and respecting their much-needed space—that will speak volumes. This can happen within our yard, with potted Plant next to lone Tree surrounded by concrete, with the sky visible above, among air, Birds, Flowers, Grasses, and Beetles. Here is where the truest support exists. Because this support is within us, too. We *feel* them. It is this support that we need to be a part of—not among life in far-off places, encroaching and infringing, but with these lives close to our homes, neighborhoods, and communities. *We need time alone with the lives that are Nature.* And words are not a part of life with individuals in Nature, or their process. Support happens when words leave to feel the breeze-filled space on the other side of the window.

Words aid in the creation of the world we are within—within each one of us as individuals, within our bodies, and our minds—words matter, even if they are nothing of Nature. Sometimes, our words become distorted when they finally leave our mouths, either by us or from others. At times, what we *meant* to express sounded much better in our heads. In dealings that are opposite to Nature, people steal moments and alter our words, attempting to manipulate our true intent based on their own, storming past our boundaries, and disregarding them for fast attempts for self-benefit. This is not who we are. Nature waits.

So Easy to Tear Down

There are moments when words misrepresent, are inaccurate, and are understood very differently than intended. We use words to hurt and to tear down others and anyone. As our main form of expression, it fails us often. We know when we walk away, when everything that has transpired does not sit well with us. We feel it. Words have an energy that is created by the words alone, which are possibly more aligned with someone's inner Nature. Even when the words used are the opposite of their motive and intent, life remains aware.

Words can paint someone in a new light. They can be positive and supportive, perhaps exactly what we need to hear, or provide what was needed in the moment. Words can be crafted to be proactive and productive, for ourselves, for our work, and for personal goals. Words can inspire, bring together, build momentum, and create hope.

Kindness exists as a lifeforce, especially in these lovely moments when words flow from our mouths—in a few words or in a few sentences only—to create moments of clarity. During those times, we are very present, speaking from a deeper place, standing in our truth, and cutting through our insecurities because they are coming from our heart. In these moments, when we speak straight, our words enter cleanly and clearly into another's home within. They land positively.

Yet, words have a lifeless side; one growing from negativity and superiority—fear.

Just as words can create a cohesive unity, words can also build walls and break relationships. Words can be manipulated in attempts to create voids and a foundation of hatred. They can misrepresent and distort, tear us apart, and build resentment. Words can remain as muck and omit our hate. Words can impact our link with others. Have we thought of how this noxious, gummy, sewage backflows into our inner state?

Sometimes, we doubt or find it difficult to hold dear how the positive words change us. Somehow, the negatives cut through us so readily. Interestingly so, too. When we are in a place where we are mindful of the failure of words, we do our best not to pass on the energy behind them to break down.

So maybe it's time for us to appreciate the profound impact words have on other lives, too.

With Nature, this process transforms so much more easily. Life in Nature is not burdened with the need for approval in the same way we are. However, *how* we use words to describe and represent life in Nature creates our self-made divide between them and us.

Perhaps a simple shift in our own self-talk of life in Nature is the seed needed to honor these beautiful relationships. Since all circles back, our human lives benefit, too.

The words we use create our thoughts, pull in our emotions, impacting the chemical release in our brains and, therefore, our bodies. The words used, possibly representing what we believe or what is good for us, do something to us physically, mentally, and emotionally—even when they are said quietly to ourselves within. Here, with proper acceptance of states of awareness, the lives in Nature step in, offering guidance without judgment.

Just as words impact human relationships in positive and negative ways, our main form of personal representation that propels our personal energy, motive, and intent—out from within and to the world—is through our voice. This delivery rides with our words. Beyond our presence, our words impact our relationships with all life in Nature. This is the beginning of the end of where we start and where life in Nature meets us.

It is our responsibility to bridge the gap, starting with our words, since we are the ones who misrepresent who we are through them. All the rest of Nature communicates through the many other shared senses. Ours have dulled from neglect, but are not gone for good. We are built with them, and they are vital to us and our use, impacting us completely. It is up to us to step back into the balance, rebuilding our inner relationship so that we might have one with those lives around us.

We use our words to create emotional distance from life in Nature. Words are used and abused to distance ourselves from our emotions. But our words are precious and important. Practice kindness with words used when referencing all the lives thriving with one another, removed from humans—starting with how we reference these lives beyond our window.

Without a Word

There are those of us who fear openness because we are afraid the vulnerability will expose us and our fears. People assume all the time with other people instead of asking for simple clarification. Do we fear one another so much that we project that fear onto life around us as a way to release it? Life in Nature does not judge or presume. Knowing this, are we able to reciprocate and act decently and justly?

Personal emotion is not part of the process. In reality, it is all about the other. Without a word, it is a pure sense of another. Can we breathe life into and develop our innate senses in such a way that they are applied to our understanding of another's experience, most importantly, with those lives in the natural world? Are we capable? Can we be present with a calm mind and heart? Can we remove any personal biases and prevent our own emotions from coloring the experience?

Through another, we experience emotion. This is what connects us all. We are enhanced and elevated beyond spoken words. It is here where Nature works silently and, when pushed, violently around us and within us to restore balance. Are we so different?

Our pursuits are not always what we think we need, and often, the focus of the pursuit is surrounded by "needs" that are more of a burden to what is true. With Nature, including those instances with special people in our lives, moments that leave us feeling balanced or uplifted are the most meaningful—when the depth of self is felt, and we feel so understood and equalized—and not a word is spoken. Reciprocate with life throughout the natural world, without words.

Our Mirrored Silence

We are not independent of our physical selves. Words impact our health. Our emotions triggered by our senses matter more to us than we may have considered. Life in Nature "hears" them clearly without a word ever spoken.

Our bodies are amazing, and reactions occur to adjust to our experiences, situations, thoughts, and emotions throughout the day, every day. This includes our inner world—the world of Nature within our bodies.

I would never claim to be an expert in our physiology, but I do know that all our parts of self support us, and they can take us down as well. We are the drivers at our wheel. Be with and learn about ourselves with life in Nature. Without a word, our within begins to respond, silently.

The energy we give off, based on our motivations and intentions, is well understood by all. It is not magic or a condition of being overly creative. Think about moments when it was known that something was wrong or someone's words did not match other perceptions experienced. Living beings are operational, and there is a lot going on within each one of us. We all give off energy. We know how reliable this is as one's representation. There might be friends, family members, and coworkers who can sense changes we have within that may or may not be reflected on the outside.

The natural world around us can sense what is going on within us. Even in discomfort, perhaps they sense our truer self and hope we might, too. When we are present with Nature, we are as transparent as can be, even with any front we fake. Our presence is more reliable than words. "Screaming" calm does not make one calmer, or anyone else feel at ease, either. Being calm radiates as smoothly as honey.

The slightest shift can be felt. If only we could capture all of the sensations and intelligence that Nature is aware of, our lives could be very different.

Through the process of feeling into the silence with the natural world close to home, we find ourselves growing in our confidence. Through this confidence of our experience with these intelligent lives throughout Nature, we find our voice—clear, honest, kind, and supportive, and without fear of who we are as individuals. We learn well from these lives that are Nature, collectively, for they are excellent teachers among us. We need them more than they need us.

Could our limitations have everything to do with our attitudes toward Nature? This is apparent in our Nature within. Each one of us is reflected in our health and happiness. Dominating limitations influence life throughout Nature, paving the way for their demise. Perhaps if we are struggling to see in a new or alternate light, our words only mirror our thinking about ourselves on some level. Perhaps this has more to do with us as individuals and in our own personal struggles than anything else. Could we listen in a way that Nature needs because it is what we need? There is nothing irrational about something as profoundly fundamental and far-reaching as emotion. For it is unspoken emotion that binds us all, including life throughout Nature. If change is what we seek, perhaps it is time for us to begin to listen.

We cannot apply words to what we do not know, *but we can know without words*. Enter: intelligent Nature.

A Whole World of Individuals

We may never be able to understand everything and everyone, but we can appreciate that, say, Ants in our home or yard are attempting to find their way like we are, and a little help can safely and effectively aid them a bit, and perhaps us, too. And if you are thinking, "Really, ants?" Imagine if more did, and the impact that would have on other populations of life that we deem more valuable and precious, including our family members, neighbors, and coworkers. It all starts with what matters most to us within our homes and our yards—within us.

There is a whole world of individual lives outside our doors communicating, learning, evolving, working together, teaching, playing, preparing for death—everyone adding to the entire community of life on Earth. We could begin to show regard by the words we choose to refer to these lives. Not because of the role they play in our own lives, health, and happiness, and not because without them, life as we have known would continue to break down. But because each one is an individual in their own right and is worthy of that acknowledgment. We would be better off with this acknowledgment. Our breath, our life-force, carries our words. We are not separate from our bodies. What we say to ourselves and of another matters greatly. Through our life-giving breath, we breathe out from within, within our choice and use of words, leaving us and reciprocating all of it back to our within place through our experiences.

Honeybees use our sun, like many life forms, like certain Beetles, to understand the direction and location of their hive. This matters when we move them to other locations. Honeybees communicate with one another in complex ways that are far too difficult for just anyone observing to understand. Individuals have created bodily movements on my hand in a way that seemed very decisive after I'd held them and allowed them to dry off in the sunshine, after they had been caught struggling in water, floating, and exhausted.

I have watched Honeybees and Bumblebees struggle from pesticide and toxin-spraying neighbors. It is painful and heartbreaking to see anyone suffer the way they do. If they did not feel pain or a pain equivalent, why the desperate, sickening struggle? Could they be screaming in a way that is different than audible sound, much like our majestic Trees? Are their screams ringing throughout Nature, with all life hearing, while falling on our own deaf ears? Maybe we might begin to think twice about spraying, even though we are uncomfortable with having Nature around us. Perhaps there is more to that statement than just words—perhaps there is more to our discomfort. Perhaps there are other ways.

Admire them and encourage their wellness, and the same might be reciprocated.

Silent Owl and Squirrel

It is possible to develop senses in such a way that they can be experienced at an enhanced or elevated level when attempting to understand another's experience in the natural world. Being completely present and in the moment is a condition, since Nature operates in the present. Our ability to completely remove any personal biases is a requirement before experiencing a higher sense or one that is fine-tuned. This may happen through multi-sensory experiences and can develop as pictures immediately or once time has passed and other functions, such as adrenaline, have had time to subside.

I came home from work and went to the yard to restock birdfeeders as part of my normal daily activities. I was met by a large female Great Horned Owl perched right above where Squirrels like to play and eat. She was not frightened by my presence as I stood in front of her less than ten feet away. I knew she had been watching me for a long time now. I felt as if I was in the presence of greatness. She looked deep into my eyes and then looked away; a memory I will carry with me through my days. She may have been resting there all day and was waking for the evening. I let her be as I moved about the yard respectfully, and after a good bit of time, she moved through a silent Owl flight to the roof of our neighbor's front yard as she stretched to start her evening.

This neighborhood consisted of single-level ranch-style homes. I joined her in the front yard and observed her from the front of our home, keeping about twenty-five feet from her. I wanted to offer her space. She would listen and move her head, grow sleepy again, and, with eyes half-closed, stretched her powerful legs, toes, and talons behind her. She began to gag and then vomited a pellet—the remains of food, fur, and bones. She did not mind my standing below her, then about eight feet away, to inspect. I backed away again to give her space.

At this point, little boy Squirrel, whom I had been watching grow, play, and learn through burying food and maneuvering around Trees and branches in our backyard, came running up large Tree, who was hit by Lightning. He ran to our roof and then over to me as if to see what

I was doing.

Mentally and very quickly in all thoughts, I tried to prepare myself for what I was about to witness. Then, he stopped. He stopped directly across from Owl; they were about ten feet from one another. He turned and looked at Owl. Owl turned and stared back at Squirrel. They locked eyes for about thirty seconds. It seems something was transpiring between them.

I have witnessed Birds interacting this way. They do so across species, and with me, too. It is something communicative, yet void of sound, and some other level of knowing that I was not privy to. I tried to understand the situation without thought or assumption and sensed no fear, no stress, no sense of space or proximity between them, and no hunting or food urges or needs. There was only trust and respect.

Owl then turned away and looked to the sky. Squirrel looked back at me and then went back to the backyard to eat some seeds. Owl stretched her powerful leg once again to display the talons used for her needs. My attention was very focused on her at that point while I refrained from staring since I have learned that no one likes this. Releasing the remains of her earlier meal, she was content. There was no need to take more than she needed. There was no need to abuse the balance. There was no need to belittle or dominate. She honored this of Squirrel, and Squirrel understood. Squirrel reciprocated. Still, there was no sound between them—only quiet.

Nature does not judge or presume. Personal emotion is not part of the process. It is all about the other; it is a pure sense of another. To ease our struggle, take a good look at how we refer to the lives around us who are seeking balance, too. It takes two to make a relationship.

Our connection to all life throughout Nature operates independently of our human language and within a whole other level—perhaps all levels. As routine as our days can be, and as boring or chaotic as life might seem to some of us, we are constantly responding to all life around us, among us, through us—through our senses, thoughts, emotions, actions, and our minds and bodies. Communication occurs even when we are not listening or jabbering away. We are *never* independent of life in Nature.

Five
The Connected Link of Empowerment

Different is Something Within

I have always liked different. *Different* made sense to me. The differences that each one of us harbors within—our truest potential—have always caught my eye.

I was different, but quietly so. I held this close. However, I was told and made to feel different (in a negative way) by those I expected to be kinder. In early attempts to share a very small part of my differences with a few I thought I could trust, I eventually found that this aspect of myself was better off protected. Quickly and early in my life, I kept this huge part of myself very close and shielded. I have for most of my life. I do to this day. But even when I was very young, I knew how to do this intuitively. *Different* is something within.

Within, a small light flickered—I wanted guidance. With and within this small light, I was reminded, time and again, that there was more to this life. I sat in this warmth and comfort, attempting to piece together what "happiness" meant.

There are ways to express *different* outwardly, but the real difference is already within the individual. Outward appearances can change, but true difference is deeper...and felt.

Being comfortable with our uniqueness does not come readily for everyone. There can be inner noise and confusion. We may not understand what we feel and don't feel because we have not taken the time to know all the good and bad inflicted upon us, swimming around and held within us—or to move on from these things. But if we start to know the natural world around us, we will know the world within us. When we begin creating relationships with Nature, our awareness and consideration extend to them. With a personal connection to positive relationships with the natural world, that link to the source of support for empowerment on our stunning Earth ceases being fractured, and starts to become connected.

As We Reverberate Out, The Natural World Reverberates Within

Nature is not a thing I need to travel to or that I can admire only from a distance, both physically and emotionally. Experiencing Nature only requires stepping outside the comforts of my home, remaining open and respectful, and personally present without any facades or assumptions others may have of me.

People are not so different from Nature. We are, after all, natural beings, too. Our needs differ little from the natural world. Yet, we have overcomplicated and oversimplified, simultaneously.

Nature and humanity must be operating together. We have broken this relationship with the rest of life on this magnificent Earth—we have broken our own support.

We have done much to interrupt this equation between our lives and life in Nature. A change within our own worlds, individually, is what we need now.

We have grown accustomed to our discomfort and stagnation, as well as our lack of empowerment. So, we try to take these personal strengths from others without any success, since that act is impossible. Instead, *the connecting support* is within each of us, to natural life all around us.

Individual life throughout the communities of Nature deserves a vast amount more from us. Our hands have caused the harm. Every little action can—and does—have a great impact, of which all of us are guilty. The good or bad falls back to us heavily, leaching around and within. The lives throughout Nature hurt because we hurt—and to them, we *are* hurt. We have lost the profound sense of respect.

We are missing these empowering relationships—physically and mentally—in our relationships with ourselves and with others, including people. But our inspiration for our lives and what that might mean is already supported by life around us. Here, inherent value belongs to wild communities and their members: Animal and Plant. Life in Nature have their own values and ways. Many parallel our ideals for which we seem to have fallen greatly. We must remain mindful, without emotional coloring by what we assume. Kind and supportive experiences speak far louder and with greater footing with a larger picture in mind...*if* we allow for such a vision to break free from limitations.

The support from life in Nature is most apparent to us when it is absent within us. Nature is empowered, while we remain discouraged. Intelligence that is balanced, kind, considerate, and one that is also our own strength—in whatever form of intelligence we may master—becomes enlivened or broken through our consideration of these relationships. With life in the natural world, we learn about support, and we are empowered. Just as we reverberate out, the natural world reverberates within.

Personal empowerment *is* support. We cannot be to the world of Nature what we are not.

Perhaps we believe we are alone because we see and experience only part of the picture that is our life. We seem to be missing that we are, in ourselves, unique, and this in itself is empowering.

For all of us, we feel broken at times, and we are unsure why. We can go about our ways with a sense of self that allows others to be, or we can be blinded by a broken understanding of our own empowerment. We can never complete or be completed by another person, regardless of how similar we might be.

Perhaps we sometimes feel misunderstood because we do not understand information that is presented to us on a daily basis about ourselves and others. At times, it feels as if it is just flowing through our veins, yet we cannot grasp it. So, we ignore it, pacify it, explain it away, blame it on others, and blame it on ourselves. For many of us, there is a return to Nature to gain understanding and perspective. For many of us, this returning clarifies a lot of how we think, feel, understand the world, and understand others—for Nature explains images, feelings, and intuition—a *knowing*.

Perhaps we often feel alone because we are not embracing life in a very natural way. We get caught in our multilayered thinking and have not included all the lives around us, expanding our perspective. When we operate with kindness and an open mind toward these individuals close to our homes and communities, always keeping *their* boundaries in sight, we begin to do so for ourselves, too.

But maybe there is a way where we don't have to give up modern elements that we have grown accustomed to, but set them aside in balance—for balance, to include balance.

And all around us, an entire world watches quietly in the background, doing all that they can to stay out of our way, unless cornered with no boundaries respected. Just like us, when cornered, fear sets in. Is it possible that we, also, are feeling cornered in a world where we only need to turn around to see that the "corner" we thought we were in is not one at all—but actually a beautiful area of life that is watching, waiting, and actively supporting one another? Perhaps this world already supports us, who we are, and who we want to be—empowering us, too. We need to take the first assured step and offer empowerment back.

Not too long ago, it would seem very unusual to not intimately know the life and areas surrounding our homes and neighborhoods. Yet, most of us are foreign invaders in our yards—so much so that we are uncomfortable just being *with* the vibrating life.

Since we *are* Nature at our core, within our very cells, would not that same fundamental support—and brilliantly vast world of individuals, the profoundly complete community of life—also empower us? If we opened our mindfulness to them, including on an emotional level, for their well-being in their own right, they might begin to trust in our watchfulness— reverberating the same back to us, with great influence on our inner health. Can we remain open enough to find out?

Lightning Grounds with Purpose

Without balance, systems begin to break down, and respect and trust are lost.

A few years back, Lightning made a visit. I am certain they were drawn to my backyard— and to me.

This bolt did not strike the top of Tree, one of the largest (if not the largest) in the neighborhood, nor did they connect to houses first. Their point of contact was right above a place where I would leave a few seeds and nuts on the opposite side of Tree (the street side). This was the only location offering a clear shot for the skies, navigating through brush and branches, dropping quite low—only six feet from the ground—moving parallel to the ground between our home and our neighbor's house. There was no other Lightning or Thunder that I was aware of during this rain.

The three of us were in the sunroom—me, Willow, and my husband. It had been raining for a good bit. Heavy rain. My husband was outside looking at the sunroom gutter, and I told him to come in and view it from where Willow and I stood. It began as soon as the door closed—his hand left the handle, and he was inside, out of the rain and water collecting on the ground.

Throughout the neighborhood, our homes were about twelve feet apart. As fast as it happened, I witnessed all of the interaction.

I happened to be facing the right direction at the right time—toward the door, toward Tree. I remember thinking how quiet the bright light was as it moved before grounding. I watched the bright white energy, which looked about six inches wide and six feet from the ground, sneak quietly between our homes. It connected with much-loved Tree, moving along Tree's bark—first up a few inches, then down a few. I watched the jump of energy move quietly to our roof toward us, then jump a second time away from Tree, heading closer in our direction to where we stood. Fast, yet quiet and still, the energy shot a third time—*even faster*—along the gutter, drawn to a grounding source. The glowing light remained bright all the way from Tree to us, silent as it moved so quickly, and yet so slowly.

As I watched, the energy pulsed brightly as it made contact with the ground about eight feet from me, with Willow at my feet, unaware. FLASH—BOOM!!! FLASH—FLASH—pulse! The smell of ozone crept in, possibly mixed with metal from the crawlspace cover where it grounded. That location that was forever changed—molecularly within, and certainly on the surface—turned very dark. There was some smoke. My husband had just been walking past that spot in the rain and puddles, attempting to view the sunroom roof.

When he was outside, it made me very uncomfortable. So I didn't ask—I told him to come in and look from within instead. His hand had left the metal screen door, and he was safely inside, about three feet from the door, when Lightning grounded six feet away at the crawlspace cover. The light from the bolt—how it moved, so quickly and so quietly—had been creeping as fast as it could toward us, following the metal gutters. The light and sound from the grounding hurt my husband's ears and terrified Willow, who began pawing at me, digging her nails into my legs for me to pick her up. My system was shocked. My vision was altered for a few days. Willow shook uncontrollably.

Even within the speed of Lightning, my senses kicked in and sharpened, and I could not help but know they were here because of me. I remember thinking this as I watched it all happen so quickly, yet it seemed to be in slow motion. At that point, I had been seeking balance for some time. This event was for the yard as well—something I had been aware of earlier that week, and each day leading up to it. In that moment, I was asking for it directly: balance. Isn't that what Lightning brings?

Memories of other visits from Lightning have since resurfaced, though never this close. This transfer of energy changed a lot in me. Balance was restored in some areas of my life. Boundaries became my priority, even if they took several years to master. There, Lightning began the process. In the moment Lightning connected with Tree, I knew why. As fast as the experience happened, it felt so slow. I had been struggling to grasp balance and boundaries—Lightning made it certain. It was time I understood.

I have the bark that blew off Tree when the bolt grounded and exploded at the initial point of contact. We collected it from the roof and the ground. I knew this individual, beautiful, old Tree would be okay—later confirmed by a licensed arborist—as we applied bark sealant to support healing. Tree's bark acts as protection, a barrier, much like our skin. I think of Tree each morning. I think of Lightning often. Although storms still make us nervous—my husband and I—we are grateful for this moment we will never forget.

Finally, with support, time, and personal work, boundaries became part of my inner world—and more visible in the natural world and in those around me. Boundaries empower and open doors. They support us and protect our certainty, our truth, and our positive core reality. They create space for others to be, and protect those who cocreate peaceful, uplifting environments in mutual trust and respect. Here, we thrive together—including all life and communities throughout the natural world. Here lives balanced empowerment—an energy force that supports and propels us forward with positivity. Empowerment seeks to ground with purpose.

When I was disempowered at my core, it was mirrored in all life around me. Feelings of relaxation, inspiration, peace, calm, and completeness arise from experiences of support from Nature—from truthful trust and walking in respect, guided with intent stemming from sound motive. These moments allow us to feel our brokenness safely. Nature steps in and guides restoration. The role of Nature is as basic as our breath.

Health Is Also Positivity

There are many times in my life that I can recall, including when I was younger, when I felt I could decipher and understand situations if I "thought" long enough. I never spoke of this, but I knew things.

Images and sensations would flow freely and line up if I calmed my brain and body and entered into silence. The process I would go through is exactly what I do today—much more developed and encompassing now, but just as accessible. In time, I learned that I was not thinking at all, but meditating—doing so in a way that opened understanding and information. I was calming the chaos and confusion associated with emotions (mine and other people's), noise, facades, and false words, allowing my senses to have space. It is very helpful, even in everyday situations.

I go to this meditative space all the time, often quickly, and it is how I understand everything when I am meant to understand. I continue to use my logical brain, certainly, but being in my heart in this way is far more reliable. In my heart space, much more is processed from seemingly outside of me, and I can move back and forth quite quickly between logic and intuition/knowingness. I believe this way of relating to the world can also take in natural processes so that, together, they work in complement, offering a more complete picture. This is a balance for me that works in all areas of my life. In time, I have learned that through and from Nature, along with who I am, I feel empowered in knowing and honoring this—who and what I am, remaining content in the now, while always wanting to become more.

Health is one of the major priorities in my life; my health is most important. I am aware that it is never separate, but flows with my relationships with life in Nature. For me, in the name of health, enjoying the present moment is a large part of my day, and the rest is focused on working toward a positive, empowering future. My goals live and flow between both worlds (presence and working toward a positive future). Positivity and creating space for what matters most generate freedom—to further support the positivity around me and within the natural world. This is first by choice and, in my own evolution, has become naturally and effortlessly second nature. This is my everyday life and world. Gratitude runs through all of this, always and without question. These are the grounding and guiding forces in my life, acknowledged each morning before I begin working through meditation. Here lives what spirituality means to me. From this space and support, laughter, joy, and appreciation flow. This is where my purpose roots and thrives.

For far too many, the chronic play of catch-up leaves us drowning, as life in Nature gets pulled into our self-induced whirlpool. In the ease of our lives, we have grown reactive instead of proactive—with our health, both mental and physical, and in so many other areas. Certainly, this extends to our treatment of life around us and throughout Nature. What will it take for us to see through the eyes of other life on this unique planet? The game of catch-up and bandages does not last.

Nature Flows When Reciprocated

It seems many of us just accept conditions and circumstances, both around us and within us, without a desire to improve. We become reluctant, feeding the reactive instead of the proactive. It is not easier to be reactive. Why would we live with something for years, justified by our refusal to face it, believing we are out of our own control? And so we go about our days and years, feeling the repercussions. Each one of us must take responsibility for ourselves.

We must take responsibility for Nature now, too. This is nothing new, although it may feel that way to many. Nature has been absent from our lives in ways that matter deeply—to our cells, our concern, our hearts, and the decisions we make. We have become disconnected from our support and understanding—not in a childish way, but in a loss of passion, purpose, spirituality, and well-being. We attempt to contain and package this world into a box when the natural world cannot be contained. In our blood, shared among all life, Nature flows—when reciprocated. This is an elegant, delicate equation, yet also a fierce force in our empowerment with all life on this exquisitely beautiful Earth.

Gratitude Grounds Us

Gratitude is powerful—more so than much else in life. In Nature, gratitude grounds us and helps us all become present. This part of ourselves needs to be restored for us to feel grounded and connected in our empowerment. To feel complete, we must do the work to be grounded in who we are—unabashed—while including our connection to life in Nature as part of that identity. Even if this means leaving them as *they* need, and often that is to leave them alone in their peace.

There is no distinction between our progress and the betterment of life for the communities that are Nature. Reactivity wears and deteriorates. Being proactive and prepared for the unknown is adaptability and survivability. But we must want what is possible, even when it is not yet clear.

It is time to step out of our comfort zones and question our ways of understanding the world around us—how our concept of self and our hearts align to include life in Nature, with or without us.

Appreciate these lives for who they are, just as we desire for ourselves. None of them should be judged by our standards. We are so similar, yes—but just like us, these individuals have a uniqueness all their own that we are capable of honoring. We can appreciate and understand all that they are in their worlds. Can we say the same of ourselves? Any struggles they have mirror our own. If it is happening to them, perhaps we need to look at ourselves, too. Nothing is ever one-sided.

Woodstock

Woodstock was a female House Finch. She was blind or almost totally blind, very determined and strong. She stood her ground. She used her voice and body language to get her point across. She used other senses, too. I was not privy to them. I do know, from response to sounds, that she knew when I was around, such as when adding water to the birdbath and filling feeders. She knew my presence in the yard. I also believe it was conveyed to her by others.

One afternoon, I was adding water to the birdbath, and she flew down because she wanted to drink. She was right next to me on the birdbath with soft, light chirps. I continued to pour.

I pushed our relationship and used my human voice to speak to her, "Hello."

The sound of my voice scared her, and she flew back up into Mountain Ash.

I don't push or test anymore after this learning experience. She was comfortable with my being there as I was moving around and about her while I filled the birdbath.

Human voice scares many lives. I find it a basic consideration to remain quiet now when among Nature.

Woodstock used my usual position in the yard as a guide. She relied on exquisite senses. When I changed the water in the birdbath, she would fly in to drink while I filled it. As I watched from inside, I followed her as she found her way using the position of Trees and the fence.

I have seen Birds help one another, both within and across species. Help is reciprocated. For my friend Woodstock, they helped her find her way. She was a fighter and would call out and stand her ground if others tried to crowd her from a perch or push her away due to their own hunger needs.

I am happy she did well within the balance we created in our yard. In fact, I think they are all doing better now—there is less frantic searching for food, just peaceful eating and drinking, spaced among branches and clean feeders. I feel this balance within me, too. Or perhaps this reflects my own mental and emotional state as much as theirs.

I closed my eyes one day and was reminded of another time I witnessed Woodstock in our yard. She had taken to our new feeder setup and was enjoying the peace of eating thistle seed. I watched her flap slowly, gaining control of her landings, as she flew from Tree she knew well—where the old feeder was—to Lightning Tree, which was the new feeder location. She tried landing on leaves and floundered a bit, but quickly popped herself up onto a branch.

There was another female at the feeder, and at that moment, it was just the two of them. I sensed Woodstock could see a little, as I watched her chirp and stretch forward toward the other female House Finch on the lowest perch. Stretching her neck, Woodstock called and called toward this female Finch. This feeder had six perches, spaced in a way that gave enough room for peace and little tension. Woodstock chirped softly to her Finch companion again and again. I watched her companion Bird hop up to the middle perch, while Woodstock leaned forward, quietly observing, leaning more to see better. The other female then hopped up to the top perch, then up again about a foot, landing on the branch next to Woodstock. Woodstock then flew accurately to the top perch and began eating.

The other female had helped her precisely assess the situation. This individual who helped Woodstock then flew back to the middle perch and joined her. They ate for about half an hour, watching me every now and then. I gave them space.

Woodstock always flew deliberately, with a few flaps, then dropping a bit, then more flaps, etc., to control where she was going since her eyesight was not her strength.

One early evening, my husband, Willow, and I were walking home from a summer stroll around dusk. As we headed back to our home, we stopped about four houses away to watch little Woodstock with her beautiful flight pattern. There she was, flying from our backyard high in the air and down the street a little to large Evergreen to roost for the night. She landed with her own unique grace near the top of this Tree. Touching down abruptly on the large, fluffy, needle-filled branch about thirty feet up, she hopped in under the cover of Tree and trunk for the night.

She was with us one summer and through fall. I did not see her after that autumn. Great warmth, appreciation, and strong affection remain. She was a sweet life that graced mine in amazing ways. This little Bird is one of the reasons this book exists. I will never forget her or those who helped her out of kindness—or their brilliance.

Six

Aware of Nature Within

Guided From Within

From a young age, I absorbed what I witnessed going on around me. I assume most of us do, in our own personal ways. I realized that I could learn a great deal from others and from situations this way—by remaining silent and observing. I was feeling much more than I was aware of. I learned to do this quietly and alone.

My early teenage years began with depth and were marked by continued instability, now riveted with shocking tragedy, time and again. Friends, classmates, friends of those I knew—individuals who were kind and extended that kindness to me—were lost far too soon. I am unaware if any of my experiences were even known by those immediately around me. There was always too much going on. I remained present for others in my life as best I could, at ages when we were perhaps unprepared to cope. But who is? I remained in the quiet comfort of having stopped sharing long ago.

There were friends, events, moments, and horrendous situations—beyond words. A lot for someone so young. And so many, over so many years, compounded. I was not alone in these moments, nor was I experiencing the worst by any stretch. Often, I was the observer. But the shockwaves I felt, along with the confusion and deep emotion, flowed from within as well as from those around me. Feeling and absorbing, I was unaware that some of what I felt may not have been my own. With each event, something shifted my world—each one imprinting itself into my memory, emotionally and physically. I watched quietly, in my aloneness. Everything, good and bad, was impacting my inner world.

I can say with confidence that it was friends who absorbed much of the shock at that time. All of us, in those places and during those years, felt it—each in our own way. Even those I did not know well—we all lived through a great deal. We tried to help one another as best we could, at ages far too young. I continued turning inward, into silence, to know.

Entering that silence became easy—second nature. I did this often, without even realizing it. I became very good at it early on, learning quickly how to retreat within. From my world and personal experiences, I began to consider what mattered in relationships—what mattered to individuals, what mattered between people—and what caused connections to break. I did this quietly and alone, protected within the comfort of my silence, keeping it all safe inside. I watched, felt, and turned inward to understand.

The years went by, building upon factors and experiences—what I felt, and then what I began to sense more deeply. I focused on the subtle indicators of connection—what brought people together and what held them there. As a result of these earlier experiences, both in my own life and in the lives of those around me, including friends—their kindness and their struggles—I started to question a lot.

Quietly and alone, I questioned the discrepancies between words spoken and what I sensed others were truly experiencing. I always kept these observations to myself. I learned early on that what many people expressed in words was not always reliable. Much could be untrue. So I found words to be inaccurate. Instead, I relied on other ways of knowing people, situations, and emotions—what was truly being felt beneath the surface...what people were feeling, truly.

Somehow, still, I often ignored the red flags in people and situations that were clearly not aligned. They were glaring and obvious. I saw them, understood them clearly, and continued to walk forward. I ignored my inner knowing.

I grew to love getting lost in this world of information and sensing. I welcomed the sensations, imagery, and emotional or physical responses that came with others, with circumstances, and with events. It was the best way I made sense of my life, of life itself, and of the world. I believe this allowed me to see things differently—to relate to others and hear them, even when nothing was said. I held it all very close. A fire grew within me. I wanted to extend this way of knowing to nonhuman life as well.

Remaining drawn to Animals and the natural world, I wanted to know how they felt and what they were thinking. There were distinct and pivotal moments from when I was very young—moments when I questioned the meaningful presence of these lives and their importance among us, asking adults who were in positions to guide me. I was told they were fundamentally different from us, in ways that did not afford them the same consideration—even at the time of death, and beyond. I understood this to mean that humans somehow held the greater place, even in that final transition. Something was not right with this—these words did not feel right to me. What of their intelligence? Surely they feel emotion, even if expressed differently. Emotions come with life. They guide us. We are not wired blankly like empty shells. Are they dismissed simply because they are different—even though they seem to know more than we do? Within, I felt a deeper clarity around these questions. So more inward I went, to listen.

In other moments, I pressed questions to adults I encountered. "How could this be?" The response was often, "That's just how it is." This sentence only created more frustration—since it, too, felt off. How someone is results that way for a reason.

I knew many people whose actions toward, and treatment of, others made them seem less deserving of respect than Squirrel, whom I would occasionally see on the wire near our home, and whom I knew in passing during infrequent moments in our backyard.

Being told something without understanding why never worked for me. This was also true when I struggled with math, and it remained true here—when trying to understand why Animals, and other-than-human life as a whole, were generally considered to matter less. If they do not matter, then what is the point of their lives? And what is the point of mine, knowing that one day they will no longer be here—around me, within my world? It felt like emptiness. And so, I continued to question—within.

Distraught, I turned to other adults, hoping for actual clarity. Again, I was given "advice"—told more than guided—and not given the opportunity to truly discuss, which was typical. But one piece of advice stayed with me and has guided me ever since: *"Don't believe everything you are told."*

And so, I continued to be guided from within.

Awareness Only Looks Different

When I was young, the world outside my bedroom window initiated a sense of curiosity. I would hear the song of Song Sparrow in the summers and loved the beautiful color and warmth in their sound. There was happiness in their voice. But for me in my youth, this voice was just that—a separate world, independent from me. I believed this world had nothing to do with me. Not that I wanted it this way, because I wanted something very different from that assumption. I wanted us to know one another.

I never knew to consider them—that I *could* consider them. In quiet moments, I would stop, quiet my thoughts, and wait for...something. Never once did I stop long enough to turn to life in Nature for much-needed stability, for strength, for my desperate need for calm, and for support. I never considered that what I was sensing would welcome me and include me if I was, within my heart, willing to be stable, centered, supportive, and calm in return, while in their company. I went within to know, but remained unaware that my inner world was more than I was capable of imagining. I walked through my days as if life around us was nothing like us. I remained in that fog. Perhaps this was what was off—what was missing. I was dismissing Nature.

I dismissed myself, hearing things in what I was told, or told in silence, with a cold, demeaning look. The unhappiness in others was somehow dumped onto and injected into me. My voice was dismissed, along with my feelings, images, and experiences. To some, I was making up a lot, and any creative side I had was used as a spear against me. Yet what I knew, I knew was reality, because I *did* have those experiences. I knew from within.

As this chipped away at me while I was trying to understand life, I began to assume there was no value in any feeling or pull from within that sought connection through understanding. As much as I wanted to be close with life in Nature, I echoed everything I was taught and told. The lives in Nature remained an "it" and were invisible to me—even in moments when I would stop to be with beautiful Tree, whom I felt so drawn to, or acknowledge a refreshing breeze full of information from lands far away, or observe Bumblebee with such focus on Flowers.

In all those years, not once did I truly see or acknowledge this Song Sparrow. He was right outside my window, and I never bothered to look again after giving up on my one and only attempt. These lives were missing from my own, completely removed. I left them there, unaware they were part of what was within me.

The fascination I had for Fireflies and Ants existed in that same breath of disregard, following all those around me in our summer days and ways to find something interesting. I discounted and dismissed them, and I never paused to wonder what they were doing or why. I could not fathom who they were on my own, even if I had tried. I never considered intelligence or unique individuals carrying their own emotions, expressed in their own way. I felt it, but ignored the evidence immediately, relying instead on commonly understood information and accepted beliefs—just as before, when I was told that their worth did not matter.

The intent behind those words still lingered within me and in the air around me. I was separate from these lives, even though I wanted nothing more than to be included within their community. They *did* matter to me, but they mattered on my own terms. In my mind, life in Nature was nothing like me and had nothing to do with my life.

I dismissed myself, and I dismissed Nature.

Living Among Them

We limit this intelligence within ourselves—the same intelligence that exists in Nature—because we limit it to those we recognize as intelligent around us. There are those of us who appreciate that the lives of Nature are intelligent. Still, we live as if we are "better than," as if Nature is nothing like us. We treat these lives and communities as resources—things—and believe they cannot compare to our own capacities and capabilities, founded on all that we feel.

In our rare and special moments, when appreciation attempts to pierce through, we may be physically present, yet not present at all. Mentally, physically, emotionally, and perhaps even spiritually, this reflects our hurt—and echoes it. It becomes a dismissal of reality—of our Nature within. This is our shortcoming. It stalls our evolution as a species, perhaps even halting it. We are not adapting to the greater community of life or to the intelligence shared freely around us—an intelligence we choose not to join, despite all that we are within.

Different is no reason for dismissal. We cannot separate ourselves from within based on differences that exist only on the surface. Difference is constant. It expands through all individual lives. Difference makes us whole—together—while sharing the same fundamental needs. We need nourishment. We need to eat. We need companionship. "We" includes all life on this extraordinary Earth. How fortunate we are to have such rich diversity. How gravely unfortunate that so much of this difference is taken for granted.

For all that we are as an intelligent species, humanity often appears limited in scope. We seem to be slipping in our everyday awareness. As we move through our lives, life around us is aware.

Caribou and Right Whales, Chickadees and Songbirds, Dragonflies, Black Bears, Oak Trees and Plants, Dung Beetles, and Bumblebees all possess abilities—both inborn and learned—that we, as humans, lack altogether. Dung Beetles, for example, have amazing abilities and remarkable skills that help them meet their needs. What takes place for each life to survive and thrive may be beyond our understanding. Inferior? No. Different? Yes. Highly skilled and advanced? Absolutely. Just as we are in certain ways—though many of our own abilities seem increasingly neglected.

Insects are not lost in this world—our world and theirs. Ants possess systems in their brains that create a world of working together, supporting cooperation within their colonies, where each individual contributes to a complex, shared intelligence. Their sophisticated communication that occurs is so much more complex and complicated—far more so than we often recognize. The communal brain of Honeybees is much like our own in that there are two hemispheres. What appears inferior could not be more wrong.

We are seen by them in ways we do not yet understand or appreciate. Butterflies have brains. These incredible lives use their abilities to determine the one kind of Plant—a species-specific Plant—that they need for their larvae to survive on during development. Butterflies use their brains, as other Insects use their own brains, to determine the right age of Plants, locations on Plants, the size of each leaf, and their ability to support the stages of Butterfly. There are so many different forms of intelligence—this *is* intelligence—adapted, refined, purposeful, and that works for the individual. Just like us.

Plants may not have brains as we define them, because such a structure would not work for them very well. Rooted in place, they are more vulnerable in certain ways because of this grounding. Instead, they have systems within, under, and around them that act like a brain. They have innate networks that function with inherent coordination and responsiveness. Imagine what might be possible if we were open to existing together—and how we might restore both Earth and ourselves.

I have learned that our internal view of ourselves often reflects what we believe about Nature. The walls we build to protect ourselves—what we shut down or hide—may mirror how we treat life around us. If we view Nature as lesser, as inferior, perhaps there is a part of us that feels the same within—a place where facades and ego cannot reach. Different is not a reason for dismissal. Perhaps we might learn from Nature's intelligence. To do so, we need to leave behind the illusions and know others and ourselves as we truly are.

We, and all life throughout Nature, are anything but unintelligent.

Know Nature on Their Terms

Termites create alternate escape routes and work together if something happens to their home, or their community is disrupted or crippled. They work with speed and purpose. With intelligence.

Ants help one another learn, and are patient, even waiting for others who fall behind. There are those who wait for new members who stop to understand a distraction along the way, before heading back along their route. Certain Ants will not leave injured members behind and help them back to their nest. Worthless? Never. They are brilliant—within their world and within their needs—and by any standard of intelligence. Yet we dismiss them, and in doing so, we dismiss ourselves and the life around us.

We convince ourselves that we are nothing like Chicken in the yard, Robin flying overhead, or Oak Tree who drops too many leaves, taking too many hours of our precious weekend time to clear. We do our best to create sterile environments around our homes, pushing Nature away. We poison, exterminate, and grow numb to the agony and suffering of others—and that comes right back to us. We leave nothing for Insects to survive beneath in winter, and little for Birds to forage when food becomes scarce. We place poison traps when we can manage in other ways, without considering the broader impact—on Owls, Hawks, waterways, soil, and ultimately, ourselves.

In our intelligence and "quick fixes," have we thought enough about what our actions and impacts have on others? We spend years creating games to play, we invest time in entertainment and distraction, yet we struggle to create solutions that truly support life. Creating torture and harm may seem easier—but is it? What does that do within us?

Birds are known to act on behalf of their mate's needs, independent of their own. They are considerate of dietary intake, favoring a diverse range of food sources based on preference or need, regardless of their own personal taste, and act accordingly. The ability of Birds and their vast senses is mind-boggling, just like all other life. Their incredible intelligence and emotion add deep, far-reaching complexity, which is a minute part of their inherent value to life on Earth as a whole living planet. Still, we disregard their needs and lives and continue to overlook them—while claiming superiority. We claim to be more.

We are all different. To keep balance on our planet, we need that difference. It is both that simple and that complex. Different does not equate unintelligent. Difference does create community intelligence. Stability. How we interpret those differences depends entirely on who we are within.

Is it so impossible to consider that we may experience life the same way? If we are not so different, could we learn from one another? And if we could learn from one another in our daily lives, might trust begin to form? And from trust, would respect begin to take root? And from respect, a sense of shared community? And if we felt that sense of community with Nature, would we not begin to care for them—and protect them? And in doing so, might we begin to feel supported ourselves? Might we feel better? Might we become better?

Are we willing to try?

Would we begin to feel for Nature? And are we most likely to protect those we deeply care for?

Assuming life in Nature is nothing like us impacts something within us as well. Just as differences exist among people and strengthen us collectively, the same is true across all life—Animals, Insects, Mammals, Reptiles, Trees, Plants—making us a much stronger whole unit because of these very differences. Like us, they are personalities housed within their own unique bodies. Our dismissal begins when we stop seeing the individual.

When we are made aware, personally, of the incredible uniqueness among each and every life form on this beautiful planet Earth, something within us shifts.

Each of us is unique in our own way and contributes something distinct—shaping and flavoring our communities, our relationships, our towns, and our world. Each one of us expresses ourselves uniquely, processes events in our lives distinctively, and experiences emotions differently. We each offer our own unique perspective.

In Nature, the uniqueness of the individual is very much alive. Our appreciation of this is a struggle for most of us. In that place of our deepest doubts and fears lives a mirrored connection to our disjointed ability to embrace Nature within. If we allow ourselves to recognize that life in Nature holds emotion and intelligence—not limited by the worst within us as people—the unfeeling, uncaring, angry, and often threatening or dismissive—we may begin to remain open, to feel the kindness in our bodies and minds through our appreciation of the uniqueness of their individuality.

There seems to be an underlying current of control and insecurity when we are unwilling to remain open. Openness allows another to be who they are, not what we feel we need them to be—by force or by words. Preying on differences as a weakness only reflects our own weakness, our fear, our insecurity. It is hard to see another's light when we are lost in our own darkness. The nonjudgmental perspective required to know another as they are becomes lost when we do not know who we are ourselves. We have all spent time in this darkness.

We allow others to define who we are based on their own needs and perspectives, often shaped by their own struggle to survive within their darkness. How we treat the natural lives around us reveals who we are and where we believe our connection ends.

Each living being carries a goodness of their own, contributing through their unique nature. If we lose that sense of worth within ourselves—if we begin to feel that "we" and "our" place is all the same—how can we know *anything* different in others? How can we allow for *anyone* else? Who we are is anything but the same. Our sameness exists in our shared capacity for good, and that reflects our motive and intent. Our place on this Earth is shaped by our experiences with individuals and how we treat them throughout Nature. Kindness and respect are our truest intentions—recognized across the larger community of life on this planet.

We are not the sole keepers of what matters most in life. Within Nature, the rapid communication of our individuality—each one of us among so many—begins to shine in a more grounded and respectful way. It becomes clear that knowing one another matters, and that emotion exists between all life. Kindness and respect for boundaries do not go unnoticed by the lives around us. Their experiences with us matter greatly.

Family can take many forms. Some species of Bats, for instance, cannot go long without food—sometimes only days, or they will die—yet they willingly share food with unrelated individuals in need.

Our views of Nature need to change. We do not need to endlessly trespass into remote places that belong—and are owed—to life in Nature. We do not need to go to Nature, to their lands and worlds, to seek something more by entering their spaces and disrupting their communities, families, and ways of life. Nature exists outside our windows, around our homes, and within our communities—ready to be experienced from a place of shared ground. We can live alongside Nature, creating space for coexistence wherever we live.

Place a potted Plant outside and care for this life as they need, through the seasons and over the years. Know Nature on their terms. Plant Flowers for Bees. Hang, clean, and maintain feeders for Birds who require and rely on seeds, not just Insects, to sustain themselves and their families along their journeys. Use alternatives. Avoid the toxic and the harmful. These substances damage and hurt them in painful deaths—not only to them, but to those who rely on them, our families, and us, to waterways when washed from rains, and to the soils seeping through to Earth. Lawns are not naturally occurring wild Grasses or ecosystems, as seen in the water required to maintain them. Consider native Flowers and Plants—indigenous to your area—for all to enjoy, and for some who need to survive.

It is within these areas of "home" that we create safe environments where we can connect and grow with these lives—physically and emotionally—together within our yards, neighborhoods, properties, and communities. Create environments where we live so that we do not feel the need to escape. Visit nature centers, environmental education centers, plant nurseries, and wild bird stores for guidance. Ask questions.

Dismissing life in Nature as nothing like us is untrue. Experiencing Nature on their terms requires stepping beyond the comforts of home—both figuratively and literally—and meeting them at our own doorsteps. They ask us to remain open and present, without any facades or expectations.

And in doing so, it becomes clear: humans are not so different from the lives that are Nature.

And something within us begins to ease—the quiet release of a struggle we have long carried inside.

Plants Are Aware

We forget our own intelligence by relying on others to tell us what intelligence is and what is required to, therefore, be considered intelligent. In the present moment and within the realities of our lives today, focusing only on our own wants and needs limits us from within and neglects the physical, mental, emotional, and spiritual aspects that make us whole. This removes us from the entirety of our complete inner world. We do not operate in a vacuum. In our desire for knowledge and guidance on intelligent thought, we neglect to include a very real and important part of who we are because we fail to consider the other side of life.

Our inability to remain open clouds any comprehension of the complexity involved. The beliefs within each one of us—that life throughout Nature is inferior to our intelligence—ripple into our own lives and everyday worlds, bleaching out the colors, fragrance, and inspiring beauty we are missing within our goals, our dreams, and our relationships with people. Our intelligence within, which is connected to all life around us, is missing the life around us. And this is affecting us—and we are doing it to ourselves.

It is very easy to fail to consider Plants within our world, let alone any intelligence they may—and do—radiate. But Plants are aware and, perhaps for some of us, chillingly so. They are aware of us and of our dismissal, too. Trees, Flowers, Plants—including those in our gardens and homes—are aware of their environments and the life around them. Plants are alert, remember, and respond to what happens to them. They are remarkable, remembering people and actions directed toward them, and relying on this information when approached again. They send energy waves into the air so other Plants can know. They communicate underground with one another and offer support when needed. Elder Trees teach younger Trees how to survive. Yet still, we remove these elder, wise Trees, and the knowledge is lost within the generation. The younger Trees do not yet know enough and struggle to survive.

Awareness may look different from how we, as people, experience situations. We even differ in how we experience one another, human to human. But awareness is still awareness. Perhaps when we interact with Plants—those in our homes, our gardens, and Trees we pass along streets, parks, and wild places—we might begin to understand motives and intentions more clearly. Intentions may be what matter most to these lives. In light of Plant intelligence, how is it that our own intelligence leaves us struggling? We could be gently connecting with these brilliant lives and learning more about ourselves from within.

Plants are also aware of our actions toward them—both kind and harmful. Perhaps they recognize kindness, just as they recognize threat. If they become aware of danger and learn who or what threatens them, perhaps they also learn who cares for them, surrounding these lives—and even individual people—in a reciprocated positive existence. Our intentions move through our bodies and follow us wherever we go. Just as Trees communicate underground, spreading information, perhaps our acts of kindness—even a loving, gentle touch—are noticed and shared as well. Perhaps what exists within us, expressed through our actions and intentions, does not remain only with us but becomes known to other lives throughout Nature.

Peanut

With little reverence offered, and despite what we have long been told, Fish do not swim in circles mindlessly or without memory. I once took home a little Fish I felt drawn to while buying food for Willow. I do not support the trade of lives in this way, and after leaving the store, I continued thinking about her for days. I had to go back. The choice was never hers. She had been forced into that life regardless of what it represented. She was the one living it. I did not want her to die in a tiny bowl on a shelf—depressed, motionless, and without spirit.

If Fish are given so little consideration by the average person, what does that mean for other life? What these lives need to truly thrive—let alone survive—we often fail to understand entirely. Could our self-made separation exist because there are few familiar facial expressions for us to recognize, especially in suffering? Just because we do not understand or recognize suffering does not make it any less painful. Tragically, these moments of fear and distress are often when we seem to notice them most.

Fish experience fear and pain regardless of what we tell ourselves for comfort. Fish play, strategize, and form bonds—even if those bonds are sometimes one-sided—with people who treat them with kindness. Yes, they are aware, and yes, they remember. Like Trees, Fish have memory, even if the process differs from ours. They remember other Fish they encountered months earlier. In the wild, perhaps they remember us, too.

I named one Fish *Peanut* because she was the color of a peanut shell. We created a little world for her, with places to hide and explore. Sweet little Peanut loved when I danced to music with Willow. She would swim toward the soft vibrations through the water with what felt like joy. I once found a sparkly decoration outside her tank that seemed to mesmerize her—perhaps she found it beautiful. After seeing her reaction, I hung it facing inward so she could swim around it, studying the sparkles from the fake jewels forming the letter "P" on a little red stocking. She seemed fascinated by it, and I would place it there daily to give her something special to enjoy.

She knew me when I approached the tank. She would swim to the front to greet me and look down at Willow, who gazed back from my arms. Peanut was beautiful and very special. She mattered to me, and I learned much from her. Although her life in a small tank was far from what she truly needed, we formed a relationship, and I tried very hard to understand what would make her life better. I tried to learn from her what she needed instead of assuming I already knew. I would go within to understand.

She touched my life deeply, and she was anything but "a dumb fish" swimming in circles, unable to feel pain, as I watched her slowly die despite our attempts to help her. She knew who I was, and she knew Willow, too. Her experiences and expressions may have differed from mine and from Willow's, but she was aware, and she was unique—like each one of us. I can say with complete sincerity, even now in adulthood, more than a decade later, that she was loved because I loved her. Intelligence and emotion create the strongest connection within.

Fish are brilliant lives with intelligence and emotion. They experience expansive senses far beyond the basics, just like us. They live in water, and their experiences differ from ours, yet they still experience life fully in their own way and in their own right. For too long, we have ignored the reality of Fish, and in doing so, ignored part of our own reality within. I learned this firsthand through my brief but meaningful time with Peanut. She deserved more than the tiny bowl she floated in lifelessly when we met. We learned through this shared awareness from within.

When we fail to recognize these lives as intelligent and worthy, we hurt alongside them. We radiate that harm outward to others. If we lack the care to consider them, what does that say about us, and how does it affect us in ways we may not yet understand? Are we missing empathy in our processes? Lacking in our own intelligence—including emotion—we become the recipients of our own actions.

Why does this matter, and why should we care? Beyond the obvious need for reciprocation in our lives, our dismissal of Nature's intelligence leaves a void within us that impacts both body and mind. We search endlessly for support, feeling the instability within ourselves and in our lives, trying to fill it with vices, places, jobs, and people, when the true home for this loss exists within our connection to Nature. We need emotional connection with life in the natural world in order to maintain health in every area of our lives. Our well-being is tied to recognizing their worth in their own right. Different does not mean unintelligent. Difference creates community intelligence. Stability. Including within ourselves.

Emotion Within

We are an interesting group. We are typically uncomfortable in our own skin when emotion is expressed, and some even belittle others for their ability to feel so readily. We tend to squirm or loathe situations when others become emotional in our company. Being comfortable with emotion, including our own, is quite different from turning our backs on how we—on either side—handle it.

Aware of our Nature within, the more emotionally intelligent we are, the better we become at understanding emotions, experiencing them, and therefore working with them to help ourselves and others, and to know others. Emotions and all of their intelligence are vital to all life.

We do not extend the same consideration to life in Nature, assuming they do not experience emotion through their own unique expressions. Because emotional communication looks different, sounds different, and is expressed differently than it is in people does not mean emotion does not exist. To assume language cannot exist in other forms, expressed through emotion, is absurd. Life in Nature has no need to write books or spend endless hours staring at devices tapped into limitless information—and even more information we merely believe we need—along with unfathomable amounts of mental "junk food." All of this affects our thoughts, emotions, and health, disrupting our intelligence and leaking into every part of our inner world. Information creates emotion.

Emotion is part of our individual experience and contributes to our physical, mental, emotional, intellectual, and mindful states. Emotions can build and support, or break down and destroy. This matters deeply within us. Life in Nature seeks those that support and nurture, and learns from what has been harmful. We, however, often become addicted to pain, inflicting it on others and, through the nature of emotion itself, inflicting that same toxicity upon ourselves within.

We choose not to "see" the learning, calculation, group-minded brilliance, and highly developed abilities behind Nature's actions and the ways lives throughout Nature support one another in their worlds—our world. We only see the final result of that internal knowledge. Unintelligent does not mean without emotion, and intelligent does not mean unemotional.

Intelligence and emotion exist on the same plane. And it is time to embrace this, even if it disrupts something within us.

Pigs are unique individuals, just like us. They are highly intelligent and possess broad emotional depth. They prefer clean, healthy environments, even though we force them into small spaces where they live in unimaginable conditions. Like most life on Earth, they have distinct personalities. Pigs are brilliantly emotional, ultra-sensitive, deeply caring toward one another, and can become sad or depressed easily. They recognize and respond to human kindness, developing relationships of their own. And they suffer.

Chickens form strong connections with one another and help each other, much like most Birds do. Geese will remain with an injured member. Roosters help raise young. I once watched Bluebirds help their female mate struggling with the size of a stick for their nestbox. The male took the stick, examined it, dropped it, flew away, and returned with one of the correct size, passing it back to the female, who carried it into the nestbox. In both emotional depth and mental intelligence, they help one another—whether on farms, in meadows, mountains, streams, or oceans. Can we honestly say the same about our general default as people?

Many believe fear is possible in Animals and other nonhuman life, but not joy or love. From what I have witnessed, this is simply untrue. It is impossible. If we experience other forms of intelligence through emotion, then the same is true for the other lives created on our beautiful planet Earth. Emotional intelligence is what much of the natural world depends upon daily. It is we who have lost touch. If we are finding ourselves out of touch with healthy emotion, we cannot assume the same is true of others, nor does it mean such intelligence is inferior—or wrong. So many people struggle with joy and the search for love that it is worth pausing to consider this discrepancy, because it impacts our health significantly. Low emotional intelligence seems to trigger frustration, fear, and, therefore, anger in many of us. Yet when life in Nature exhibits fear, it is usually for a sound reason. Can we say the same of our anger—or even our rage?

Those who work closely with Animals have stated that they do not experience anger coming from them. They do, however, witness creativity, fear, shame—or a sense of self-awareness—and appreciation for beauty within their experiences. They witness companionship, joy, excitement, support, and love. Animals, Birds, and other lives reason through memories formed from emotional experiences that help them survive and thrive. When they are free—their true freedom—they remember locations, individuals, weather patterns, food caches, and how to construct homes and safe places to care for their young. They must learn quickly and remain alert, strengthening their intelligence. They cannot sit for hours staring at screens as I am doing at this very moment. They cannot endlessly scroll through images all day long as so many people do. They want to live their lives—not the lives we force upon them or push them into. We need to be part of that life, too. We are designed for it physically, biologically, chemically, and emotionally. We need life in Nature.

We do not like to accept accountability for the ways we have created life up to this point. Because of this, we fail to see clearly moving forward, holding a different and more supportive approach within our hearts as we live our lives, experience human connection, and embrace our relationship with life in Nature around our homes, neighborhoods, and communities. We remain separate within our own healing and within our interpersonal relationships.

We live in a world where emotions are treated as a weakness, where being emotional and feeling deeply with and through other lives is considered wrong. A world where emotion is falsely defined only as uncontrolled crying, fits of anger, or rage—a reflection of deeper struggle. Many of us are so uncomfortable when another person expresses heartfelt emotion that when it happens, we are left with no response and no meaningful connection in return. Yet there are so many emotions that empower, inspire, and offer hope. Emotions simply make many people uncomfortable.

Still, we claim emotions as ours alone. We tell ourselves that only humans experience emotion. Other beings—Animals, Plants, Insects, even Song Sparrow singing outside your window—are reduced to "things" operating only on instinct. An "it." We hold emotion in high regard while refusing to acknowledge its expression and mastery in other forms of life. We remain unaware of the magnitude our other senses have on us and on the way we relate to the world. We assume life in Nature is nothing like our "brilliant" selves, yet we continue struggling to make sense of our lives and our world. Even when we believe all the answers exist within the downward stare at our fingertips, we look up only to miss what is directly in front of us.

When understanding and awareness come together to help interpret our world, there is no place left for dismissing life as unintelligent or unemotional. That dismissal feeds a world of "us" and "them"—a world where individual lives are reduced into a cold, heartless mass, supposedly void of feeling and intelligence: an "it." We avoid the root of our struggle—we dismiss Nature around us and within us. We have created complexities that have muddied what once came naturally to so many of us. It becomes confusing, and some people can no longer bear to feel it.

Nature mirrors us in this way: kindness and consideration are reflected back naturally, creating a safe place for us—a sense of home. Health flows back to us through our minds and bodies, offering a strength unlike anything else. But when we destroy that connection, the damage also returns to us through the chemicals of our thoughts, brains, and bodies. Are we afraid that if we truly accept their intelligence, the door to understanding their emotions may begin to open, too?

Childish

Any sort of deeper connection with Nature and the life that is Nature is often considered naïve or childish. Any connection or observance is labeled "cute"—an almost syrupy attachment viewed not as admirable, but as inexperienced and naïve. With hypercritical minds, we remain unaware of all there is to appreciate in this relationship. Yet having the foundational support of healthy relationships with Nature would only enhance and expand our desire for supportive relationships with other people. Nature is of this Earth. The lives that are Nature are as distinct and unique as each one of us. Together, great good could come from acknowledging our shared needs, and there is nothing childish about that. Still, it—we—and they are ignored.

It is easy to fall back into the belief that affection for, and relationships with, life in Nature are irrelevant. That emotional attachment to Nature is somehow childlike. Yet our relationships with Nature are the ones that need our attention most, because they are our core, our foundation, our grounding from which all other healthy relationships evolve. Relationships with natural beings and life empower and support human relationships. But in order to have this, our inner Nature must be open and ready for it. Still, for the most part, Nature is viewed as something only for the childish or inexperienced.

Our ability to intrude upon, ignore, and remain painfully unaware of the space, needs, and suffering of life in the natural world is something we have mastered—and it has caught up with us. It has become a relationship of struggle and pain that is impossible to ignore once it begins affecting us within, as it already has. The hurt and pain we inflict upon those who are, by design, the source and inspiration of all that is sustaining and good keeps our own pain steady, even when ignored. Our inner Nature witnesses this and is part of it. We are the life we have created through our treatment of others. Our actions impact every part of who we are, rippling outward into our worlds each day.

I have found that Spiders trapped inside my home are often desperate when discovered in sinks and tubs. Without a proper food source, they usually die of thirst after failing to find food. I gently capture them with a cup and paper to keep them safe. I place a few drops of water nearby, and 99% run to it and drink—for a long time. Then I release them outside so these amazing creatures can continue living their magical lives. They provide so much good around our homes and yards. They are ingenious engineers and craftsmen. They help create balance among Insects, Bugs, Birds, Mammals, Reptiles, Amphibians, and all the others I may have neglected to mention—including our friends, brother and sister Fish.

Once upon a time, I was not very fond of Spiders. I never disliked them, but the thought of one on me felt like minor torture, even if it was only a taught, learned, knee-jerk reaction. The worst moments were when one became trapped inside my car, trying to make sense of this strange world called a vehicle, while I drove and they hovered above my head. Perhaps they were sending representatives to get my attention. It worked.

I began watching them and learning about them. I observed two Spiders working side by side with webs at my doorway, perfectly positioned out of reach from anyone entering the home. I watched them face one another, four legs touching, almost as if they were having a meeting or conversation. Perhaps they were enjoying each other's company. Maybe they were trying to determine web placements that would benefit them both. Or perhaps they were simply resting together, sharing a quiet moment. They were communicating, even if their communication was never meant for me to fully understand. I respected this, and in return, they respected my space as well.

I once had a particular Spider living in my apartment all winter because it was too cold to release her outside. I watched her struggle near the end when food became scarce. I felt a deep loss when she died. During that harsh winter, I moved between work and home through what felt like a frozen tundra beyond my door. She was there, and she fascinated me. She slept in her web above and to the right of me in a corner that belonged to her. She would shake her web whenever I crossed too closely into her trusted space. She watched me. She rebuilt when needed. I learned a great deal from her. Along with my sweet Dog Willow and my Houseplants, this little Spider became another companion, keeping me company in a new state where I knew very few people.

It is okay not to want a Spider crawling on us. I do not think they particularly want to be on us, either. Knowing they experience fear as well, there is likely a distance they feel comfortable maintaining, too. We can kill, or we can help. It is a choice. In helping, an energy begins to transfer when we act supportively, just as it does when we destroy. Through this support, I have now experienced moments where it almost seems as though a trapped Spider comes to me for help—though these moments are rare. Perhaps they always were, and I simply notice now. I catch them, offer water, and after they drink, I release them outside.

A trust between us has been established. So few interactions happen now that it feels almost like an understanding has formed between us. Ignored? No. But it was not always this way in our relationship. Realizing that we both require water to survive was a profound step for me. The fact that they could sense where water existed when trapped inside walls seemed brilliant. In turn, I sensed their desperation and acted. Honestly, if I relied only on thought, I would likely ignore them and continue focusing my attention elsewhere, avoiding inconvenience or disruption. Acknowledgment has become a conscious choice.

One Spider trapped in a very hot closet and close to death from lack of food—and therefore moisture—drank for several minutes. As I carefully moved them outside, the little one continued drinking. I waited patiently before placing them within a garden Plant. In the final moments before disappearing beneath the leaves, I gently placed my finger beside them, and softly, two little legs lifted and rested against me in return.

Many of us appreciate the beauty of our natural surroundings. We admire and enjoy moments during hikes and walks, or with potted Plants on our porches, in our yards, and gardens. In those moments, when we truly pause to enjoy Nature, we experience the fullness of the present. Yet most of us limit ourselves to only a handful of such moments each month, if we are lucky. In flashes, something awakens within us—something that makes us feel connected, appreciative, and perhaps inspired by the lives creating these beautiful surroundings. Those moments are not lost; they are carried within us, held close and remembered. Nature watches quietly.

"Getting away" implies leaving a place or way of being for a period of time. Many of us treasure those special moments on vacations or getaways when we experience Nature deeply. It is difficult to imagine a life where this connection is not occasional, but woven into every day—where we experience Nature positively and supportively as part of our normal existence. In such a life, a "getaway" would instead mean those moments when we become more removed from Nature, mentally or physically, and returning home would mean returning to completeness through the relationships we have built with care.

By failing to allow Nature to be part of our basic sense of self and self-worth, we ignore Nature's importance and value. When we are not active participants in acknowledging the core connection between ourselves and life in Nature, our experience of Nature remains limited. For most of us, these lives are not truly part of our daily lives in a lasting way. Nature—and our own nature—has been ignored to the point of unimaginable consequence. We are struggling to stay afloat.

Are We Worth It?

Our inner world, our inner Nature, includes the landscape of our relationship with Earth reflected within each one of us. This inner world encompasses much of who we are—our attitudes, beliefs, mental health, and emotional health. It is all that we are, including our intent and motivation. Inner Nature also includes our physical workings. They are all connected. Each is impacted by our attitudes, beliefs, mental health, and emotional health. Here lives our strongest support and deepest connection to life in the natural world, too.

Yet it is this very place within that can either change our course or continue to hurt us. In our daily lives and inner worlds, a void sits quietly, always searching for the outer connection of acknowledgment. Among the chemical reactions occurring within our brains as a result of our emotions and experiences with the world around us, our dismissal of Nature—and the internal toxicity created by that dismissal—spreads through us, through our inner Nature, beginning with our actions and experiences. Doing without this connection, we feel the absence in some way. Are we allowing ourselves to feel worthy of our relationship with Nature within?

Life throughout Nature, collectively, is a fundamental support of our health and happiness within each one of us. Without opening ourselves respectfully to all that Nature and the natural world truly are, we limit our experience of life and rob ourselves of positive, supportive, life-changing experiences. In our hurt, toxic thoughts create very real chemicals stored in the body through experience. Perhaps within our senses, we are always searching for supportive, connecting experiences with individual lives throughout Nature. Yet we assume this support does not exist and instead attempt to replace that need with human-created substitutes that our bodies and inner state ultimately reject.

We are not independent of our bodies. Who and what we are contribute greatly to our physical selves. Our inner Nature has tremendous influence on our physical health. Toxic thoughts alter our biological functioning through processes activated by the brain. The memories created within us—and the experiences through which those memories are formed—release chemicals from our brains that can either support us or become toxic responses within the body. Our thoughts build our memories. Our thoughts influence our memories, whether good or bad, impacting our cells—our inner Nature. Experiences trigger emotions, emotions lead to thoughts, and thoughts form memories. Our experiences become linked to individuals and situations through our senses—and we have many senses, just like the life around us. It is here that we must mature and look inward to understand what our fear truly represents. Nature watches quietly, waiting.

Our positive moments and thoughts of sitting with Bumblebees on Flowers—feeling that wonderful pull when they buzz close to inspect us out of curiosity or for reasons entirely their own—create beautiful experiences. The senses trigger feelings, feelings become thoughts, and thoughts become memories. The opposite is true as well: If we were stung by another kind of Bee as a child, perhaps after swatting and flailing in fear, we may have pushed our fear onto them, causing them to defend themselves in fear in return. Experiences formed through fear can also create lasting, toxic memories. Our experiences impact our health in profound ways. And life throughout Nature seems to be aware of this. We can change the thoughts formed from our experiences and everything that follows, allowing for a shift in our inner climate. But to do so, our attitudes and outlook must take center stage. Like any relationship, these are not one-sided. Nature functions relationally. Is our sense of worth worthy of this relationship?

When we dismiss the beauty in each individual life throughout Nature and replace it with a dollar value, we dismiss their worth. Their value becomes tied to a human-created currency. There comes a point when we must look honestly into the mirror of our place within the world and examine our motivations. We hold closest what we care for from our core—from our heart. If we feel deeply connected to something, someone, or somewhere, we are far more likely to hold that connection sacred beyond financial value.

Many of us appreciate Nature. We love being outdoors and visiting beautiful places. We love gardens or sitting quietly within them. We love oceans, mountains, the warmth of sunshine on our faces, and the scent of Trees and rain. We experience moments of acknowledgment and appreciation. But often, these moments are one-sided. For many of us, interaction with life in Nature occurs only in artificial situations, far removed from home. We observe lives held captive in unnatural conditions. Here, depressed and mentally unwell, we limit our focus on what is happening to them because it becomes difficult for us to truly feel. When something horrific occurs, that horror lingers in the air we all breathe. Even in our dismissal—whether intentional or unconscious—Nature still runs through our blood in ways many people refuse to acknowledge. Refusing to admit it does not make it any less true. Nature exists within our blood cells and is meant to nourish us, yet we fail to appreciate this, and our actions reverberate back into us. The effects of our current condition are evident in our health, our happiness, and the state of relationships around us. The struggles of life throughout Nature are mirrored within us—in what we believe and what we know to be real. We have lost inspiration. Can we pause and honestly ask whether we are worthy of a second look?

I have been fortunate enough to witness White-breasted Nuthatches teaching their young how to maneuver along the bark and trunks of large Trees. The little ones struggle at first, tiny wings flapping as they slide downward, their small toes grasping to cling to bark. Parent Birds remain nearby on branches, offering guidance and support—information—through vocalizations and attentive watching. I have seen mother Squirrels teach their babies where to find water, food, and safe places, with young ones clinging to their backs and peering over their shoulders while learning. I have watched mothers touch noses gently with nervous young Squirrels, conveying information, comfort, reassurance, and love. Young Squirrels can be incredibly attentive while playing joyfully with siblings, Rabbits, or young Crows. They can be remarkably Cat-like in their bursts of energy and playfulness—much like Fawns and young Deer. I have had young Robins, Finches, Rabbits, and many others left peacefully in our yard while parents rested nearby or flew off briefly. Sometimes they clearly knew an adult Crow was perched high in Tree, standing watch over their own young family member in our yard. Through communication I could not understand, everyone remained safe. The sense of play is the same. Learning is certainly present. Social development, absolutely. Joy and love—without question. Do we allow ourselves to feel equally worthy?

Are we struggling because we believe technology alone is the answer for what we naturally lack? Have we neglected our own progress as a species among millions of others? We seem excluded from the larger conversation because of our attachment to fear, control, insecurity, and words that leave us empty while creating even more fear and insecurity. Why are emotions so closely tied to communication, and why do we use words to explain away and distance ourselves from that which naturally connects all life in healthy and supportive ways? We need communication with one another. We become far more whole when we can set our defenses aside and simply be with life around us in ways that nourish body, mind, and spirit.

Taking a moment to respectfully observe the wisdom of Insects building homes, finding food, and communicating with one another bridges something within us. Animals observe and learn from one another to better care for their young, teaching each other about food, environments, Plants, and safe places. Plants and Trees respond to rainwater far differently than they do to the water I provide, taking in signals from the breeze and wind just as Animals, Birds, and Insects do. I am fully aware that I know very little about the vast world to which I am not privy simply because of who and what I am as part of humanity. The information and support traveling beneath our feet, through the air, between individuals and across species is so profound, immense, and interconnected that we can barely comprehend it as we exist today. We continue viewing the lives of the natural world as inferior—especially in intelligence and emotion—because it is easier for us to do so. Yet our bodies know otherwise.

I love learning from Birds, and they have taught me a great deal about who they are individually, how they function in larger groups across species, and how understandings exist among themselves, Hawks, Crows, and Owls. Through the incredible insight they have extended to me from within their worlds, I have learned tenfold about myself, my values, my belonging, and my purpose. Above all else, as profound as it may sound, I simply enjoy having them near my yard and the places where I live—places where they can eat, bathe, relax, learn, teach, play, and simply be. So much of who I am today, and what I have accomplished, is a direct result of the intelligence and support offered to me by these lives and the many other lives I have been honored to know through brief encounters and years of shared presence across distances and locations. Here, with them, the support needed deep within—emotionally, intellectually, spiritually, mindfully—is offered and shared in a fuller world and in a fuller way. The picture of life, and of our lives together, is made up of unique and mighty individuals.

Do we care enough about ourselves to find the lives around us worthy of our kind consideration? We must first feel that worth within ourselves. Nature leans forward...listening.

Seven

Inner Intelligence, Empowering Purpose

Purpose That Seeks Good

From the time I was in middle school, I began questioning *purpose*. As time went on and I entered college, I searched for what this might mean for me. After college, I continued searching. Through the start of my career and into graduate school, I remained searching still. By this point, keeping any deeper interests that mattered to me hidden and protected within had become second nature. Quietly, these thoughts filled my daydreams, and I found great pleasure in getting lost in what purpose meant for me; what that purpose felt like, and what images would flood my mind. Piecing and weaving information together from real-life experiences, seeing if it felt right, undoing and weaving again. Quietly, this drive evolved into a need.

I was following a path that was not always clear. I do not believe our paths are intended to be clear all the time. How else would we learn? What was clear was that I was guided from within and through experiences. I was guided by life outside of me and within me—all in unison. There were moments and periods when I would attempt to ignore that inner guidance or pass it all off as imagination, following the pattern of what I had been told early on. But the moments of knowing and the opportunities in life would arise again at key points, and I felt pulled—again and again and again. Though unclear to many around me, the choices I made and the steps I took had direction and were steering me somewhere. I just knew...a knowing. I kept it safe.

This inner and outer support was everything to me. It was out of my control and was not present with everything. I could not call it into being. If I sensed nothing, then that was what was afforded to me. Knowing happens when it is meant to, but most do not listen, and it fades early on. Perhaps it never existed for some. Perhaps it is simply stronger in others. I believe it is built within naturally for some and cannot be created. Perhaps for others, it expands when utilized, like a muscle growing stronger with use. I always knew when it was time to listen.

Something happened in my early to mid-twenties, and that knowing and guidance stopped. I had found my own apartment and space as soon as I could. Nothing especially remarkable, but very reasonably priced and mine. I knew I wanted to do something meaningful with my life, and I knew I wanted a happy home. But all the other areas of my life felt far less guided from within. And for the most part, I ignored them. I ignored so many red flags when it came to people and situations in every area of life. Everything became murky. This left me unknowing. It was as if I had ignored the knowing and what I sensed for so long that it simply turned off, and a different way would now be available to me because I was not understanding. Perhaps, and more likely, it was meant to come through differently. I needed to learn more about myself.

Unaware of this at the time, I began relying too heavily on thinking. I began relying too much on what others were saying to me—sometimes from people who truly knew very little about me.

I wanted to pursue purpose and only purpose. Instead, I began following the paths of others—which is wonderful if that is what someone truly wants and needs. But I knew from a very young age what I wanted and needed: purpose and a happy home. I wanted to do something in my life that felt right to me, something that helped the world in some way, and then come home to quiet and calm. No chaos. No uncertainty. Peace. Instead, I became lost and no longer knew what I was doing.

Sometimes, in the moments that matter most, we leave ourselves wide open and allow others to take our power and our voice—the very voice needed to fuel our purpose. Even when those moments are well-intended. All of us speak from our own experiences. Still, whether well-intended or not, other people can begin painting our picture for us. For me, I did not like this. In moments when I became caught in another person's story of me, my voice turned harsh, sour, and unlike myself, even though it was still my voice in that moment. I allowed it. I thought too much and followed paths that made others happy, but not me. And I knew it, too. I was far from loving myself and began to no longer recognize who I was in the mirror. We become vulnerable to the states of others when we lose our sense of self-respect.

Trying to calm my mind and enter the silence, as I had always done to know what to do and how to proceed, only left me standing in silence itself—as if in a room of white walls, no windows, no color, no sense of anything except empty words floating around me. Nothing came through. I had no inner guidance. What I had relied upon my entire life to understand situations around me had suddenly stopped, and it made me feel blind to everything around me. I felt empty, without information, lost in a fog, and unable to see. I replaced feeling, sensing, and knowing with thinking. It was a terrible place to be. Looking back, I did not trust what I already knew within myself, even when I knew what to do and chose the opposite decisions that halted me. Being caught in situations can feel like drowning when we are not supported by our own answers to questions never even asked of us.

My purpose felt deflated, wanting only breath. Ignoring information had finally caught up with me. I lost focus on what mattered most to me, and something inside shifted. I see the signs now—I was lost.

I remained stuck in that tide pool, with a current spinning me around and around for years. I lost track of my purpose, and I began slipping into unhappiness, rarely thinking to connect with life in Nature. The unhappiness took over. My unhappiness, misdirection, and absence of inspiration caused a flood of negative emotions along with poor decisions. My roots of support were neglected and decaying.

My purpose was my center—at least it always had been, even if not fully developed. Purpose was always evolving, guiding, and continuing. Purpose inspired me. Yet during this time, I followed the path while ignoring every warning sign, without my center and support. I lost my sense of worthiness for love. That is often our sign to change our focus—and quickly. Love needs space within us in order to exist.

I believe I needed a new way, an additional way, to understand and know. Somewhere, it was understood that I still had more learning to do. The pause in knowing and in tuning into signs and situations became necessary. Here, within the confusion and the loss of purpose, circumstances began revealing where I needed to heal and evolve my inner intelligence.

Thoughts on purpose: it began as a swirl within my mind and throughout my body, shifting and turning through events in my life, always directing me back to myself—to my life. A mirror. There was a need to do something more than what I was doing. And it truly was a need. Nature was always watching, closely and from afar, waiting.

I am no one especially remarkable, yet we are all special in our own ways. I do know that purpose matters, though in different degrees and volumes for each of us. The varying pull matters deeply to every individual. Judgment is not found within the unpolluted flow of purpose. Neither are negative or underhanded motives; those always rise visibly to the surface regardless of what manipulators may believe. Negativity always flows back to those polluting the waters, polluting within themselves. Purpose that seeks good for a larger community through help and support becomes an encouragement greater than any one individual. Purpose is empowerment and light.

Evading Inspiration

When I lost my inner guidance and inner knowing in my twenties, I allowed what other people needed or wanted for themselves to be applied to me and to my life, even when I expressed my unhappiness. I became uninspired within my heart. Others who were unwell targeted me.

Years later, I understand that I was confusing what I sensed in others and claiming it as my own—even the unhappiness, anxiety, and anger. I had no protection within, radiating outward around all I cared about. I was unaware of it at the time. It is easy to remain unaware of this when surrounded by it. I was completely unaware of how to separate what I felt and needed from what other people needed and what others were feeling. Lost, I claimed much of it as my own. The support I truly needed was found only through developing my life alongside those lives in Nature.

Without these fundamental relationships, our relationship with ourselves can become clouded, and our boundaries with others become much weaker—often missing altogether—than what we need in order to remain balanced. For me, without these honest relationships, I developed a misperception of purpose, and that impacted my own personal greatness and what I was meant to offer: my unique brilliance.

Our malnourished inner intelligence can become a void that is filled by following along in uncertain areas of our lives, in major life decisions, and in attempts to make other people happy.

Still, my inner Nature held on with that never-ending flicker within that we all possess. Without any real capacity to connect, the cracks between myself and my true world, my inner intelligence, and my passion, began to leak in small drips, and my relationship with life in Nature drank in those drips deeply. Still, I did nothing to support these relationships and ignored my own emotional health and balance along the way.

I accepted the life I had created, thinking there was nothing I could do to make it better. It is difficult to feel inspired when trust leaves our lives, leaving us uncertain even of our own inner knowing, aware that so much feels "off" and deceitful, questioning any sense of personal control, positivity, and support—questioning our worth.

I had mirrored this lack of trust with Nature that I was experiencing within myself. I wanted to feel and breathe a sense of personal greatness and follow inspiration toward purpose. In reality, I felt the opposite, as though I were bleeding out any brilliance I had to offer the world.

A mistake can only be corrected when we acknowledge it and receive the support needed to make it right. The strongest and most reliable support—support that does not judge—is always available through Nature. We are not separate from this source of support. Yet I did not trust in it and instead turned all of my thoughts inward, relying only on myself.

A sense of purpose cannot fully flourish without a sense of our own inherent greatness. This is something felt deeply and known within. When we lose our connection to the greatness reflected through life in Nature, we begin to believe we are insignificant or irrelevant.

Bleeding Brilliance

During this time, I could not move forward in a positive trajectory in this world with any accomplishment that truly mattered to me. I lacked focus, and I lacked it within as well. At times, I lacked calm, balance, presence, and kindness. We are all guilty of this. Our grounded calm within our inner Nature is necessary in life. Nature seems to understand this more than people do at times. It is more felt than not.

It seems that as we age and our focus shifts, we begin to appreciate times when connections with life felt more present and real. Many of us grew up playing outside for extended periods of time, often every single day. And we played—sometimes all day long. We watched some television, but it was more fun to play outside with friends or get lost in our imaginations alone. We solved problems, created ideas, and used creativity to produce desired outcomes or simply to see what would happen. Games were more board and less electronic. They were physical and interactive, shared with others in the present moment. We supported teammates when they needed encouragement from the group. We rode our bikes and went home when the sun was setting, when we heard a bell, or when the streetlights turned on just before dark.

We were not held to tight schedules. Competition stayed mostly on the playing field and between teams, not parents. We learned grace in losing and grace in winning. We understood what it meant to play fairly. There was time to discover, imagine, and dream.

We dreamed of ideas that had less to do with money and more to do with creating, building, and understanding one another and the world. We would lose track of what day it was during the summer and run home only when we needed a drink of water—if we even did that. Sometimes, a garden hose attached to a random house worked just as well. We took only what we needed, making sure to turn off the faucet afterward, grateful for another person's willingness to share. And they shared.

We breathed outside air among Trees and Birds—even if only a few—along with Squirrels, Creeks, Forests, and Streams, if we were lucky. We felt natural light, sun, and rain, perhaps snow, depending on where we lived and the time of year. We interacted with one another—our friends, maybe our friends' parents, and if we were truly lucky, our own parents or siblings. Our phones remained inside our homes, anchored to the walls. Computers were mostly found in offices, if at all, and generally stayed there. More often, jobs ended at a certain time of day, allowing us to focus on other parts of our lives. And we could focus. And we had connections.

Often, this world included more time in Nature. Not always in trusting ways, and not always through intentionally seeking Nature out, but we connected more simply by being in situations where Nature happened to be present too, even if quietly in the background—a place many of these lives seem to prefer. Still, we were among greatness.

So many of us struggle because we are without inspiration. Without purpose. Purpose blooms from the soil of inspiration. There is a strain within us for what provides encouragement, focus, and attention during those moments when we wonder if there is more to what we are doing with our lives—more than what exists in our present circumstances. In those moments, if only we could see the parallels between our need for inspiration and our need for life in Nature. Not always in grand moments of awe, though those are certainly possible too. Inspiration can come through the passing energy of brilliant Bumblebee. Without life in Nature held within our hearts in any real and meaningful way, our brilliance begins to bleed away. The rush to "fix" this loss only adds to our struggle, pushing away the greater good of all life.

There is a fundamental void in many of our lives that we attempt to mend with things, people, food, approval, self-medication, and mood-altering states. Life should be enjoyed, but stuffing fillers into places meant for something deeper will never provide lasting relief for the completeness we truly need. That completeness resides within us. Things, people, food, approval, self-medication, and altered states only cloud us. Our relationships with Nature affect us in ways we have struggled to accept, but when we finally do, the change is lasting. These relationships can never be replicated through technology. Without this source of support offering guidance in intelligent and physically healing ways, our understanding of our complete picture—including our purpose—begins hemorrhaging inside us, just as the struggles of Nature become visible all around us. Beginning closest to our homes, we seek inspiration while walking past the very lives whose brilliance we are bleeding away.

When we try to follow another person's map for our sense of self, inspiration, or purpose, we may lose sight of our own legitimacy and the loyalty we owe ourselves. We are all unique, and yet every one of us—including life throughout Nature—shares basic needs for survival. We are alike in that we all fulfill those needs in different ways. I believe purpose is itself a need and that it emerges from our need for Nature. Any focus that is positive and intended for the benefit of others extends positivity outward into the world and back into ourselves, and in turn we are supported by Nature. How, when, and in what form those needs are met varies greatly. Some of us are fortunate to receive support through friends, family, and coworkers. Not everyone is so fortunate. Feel respectfully with Nature in order to know within.

The truest and clearest support for all of us is the kind that reaches our core and our cells, creating health and balance. We can only offer to others what we ourselves are within. When we are unaware of this—or worse, knowingly violate it—the complete support needed for our greatness and our purpose is lost. We bleed out our brilliance. The grounding, honest, and unwavering support of Nature becomes our personal map and our footing through life.

These relationships are not ours alone for the taking. Relationships are never one-sided. They cannot be fabricated. That is impossible. Nature becomes a source of encouragement that fuels inspiration when the relationship is mutually supported. These are not superficial or shallow forms of inspiration, but real exchanges and genuine connections. They flow to us from within.

There is an energy to connections with life in Nature. Our self-protection of facades fall away, and our real selves step forward, sometimes without our awareness, and our true needs become clear—often differing from what we thought we needed. Life in Nature sees into our center and cuts through everything we try to hide. Here, we feel safe because somewhere deep within, we intuitively know we are among sisterhood and brotherhood. This is not through fear or intimidation, which may belong to the workings of people, but through balance and support so that we may realign with all that we are and all that we care about. Through this state, through Nature, inspiration finds its way into our world and, if we are willing, can guide us toward achievement while supporting our beliefs, hopes, and dreams when aligned with positivity. Again, this relationship is not one-sided. Our beautiful and rare planet Earth seems to know this as the only way.

Nature is inspiring. Without Nature in our lives in reciprocated and encouraging ways, we fail to see the magic and everyday brilliance echoing throughout life and within ourselves. Devotion and focus on family are demonstrated by Crow. The love and joy shared between a mother and father House Wren while teaching their babies where to find food and navigate their world is deeply moving. I remain endlessly inspired by how quickly Animals learn and how brilliant Insects are at what they do, even when it comes to lives that may fall short of our perspective. We yearn for connection to their world. In Nature, purpose becomes clearer. The desire runs so deep because it has always existed within us.

Deemed Irrelevant

Facades and guards we have manufactured to help us traverse life may be necessary when dealing with certain situations that do not provide an environment for us to thrive. Nature operates without masks, and they become irrelevant when we spend time with life in Nature. In Nature, we are known.

When we are preoccupied, focusing on what we believe we need or think we need, or simply thinking too much, we are not remaining present and open to "hearing" life in Nature. With anger and hate in our hearts, minds, and actions, we lose the greatness of having purpose. Believed to be immaterial or unconnected to our daily lives and our whole being, life in Nature is abused. Nature is deemed irrelevant even though we are Nature within our core. Unsupported in what these lives truly need, not what we think they need, they are left for dead—including within us. Without these lives held within our hearts, we continue on a path evading ourselves and remaining lost without purpose.

The source of our inner intelligence guiding purpose is in our motive and intent—what is in our heart. Who we are is most apparent with natural life. The choice to have a healthy relationship with the lives that are Nature must be real, deep, and genuine. Masks melt away to reveal our innate brightness. We need to offer trust in return. Our motive and intent are known among Trees and Birds.

Our intentions need to vibrate from a healthy inner motive for support from these lives to exist. When our heart lacks openness and willingness to "hear" another, our balance is thrown off. Life in Nature is anything but easy. Through it all, Nature remains the master of balance, and the individual lives in Nature respect and are aware of this within one another. We, as people—as humans—are often the masters of imbalance. Nature is wiser.

Trust impacts all areas of our lives. Trust begins and ends with our relationship with ourselves and sprouts through our intentions and motivations in life. Self-trust opens doors to confidence that is sensed by those around us. Remaining trustworthy blooms inspiration and purpose because, with Nature, trust is heard.

Purpose varies between people. Nature-supported purpose always remains linked to the rationale and good of the whole. And when our purpose is to help people in positive ways, we help life in Nature, too, when we are trusted because we are trustworthy. We need life in Nature to be respected, just as we need it ourselves.

There are times when our logical intellect pushes what we think we need, like it did for me so long ago. But even with basic logic, life in Nature has always stood front and center. As we learn to utilize other senses and other forms of intelligence, we are provided with very clear support and direction that work alongside learned intellect and logic. We cannot help ourselves unless we truly want to help ourselves. We cannot be balanced with Nature and live in greatness of purpose unless we genuinely want it. We need to meet these lives in Nature on their terms, too. Perhaps it is time to step aside and ask whether we are worthy of their trust. What is our inner purpose?

I have been so determined to help Nature in some way that felt right to me that I disrupted my own connection to this drive. I tried to explain this away—I do not have the money, or I am too busy with work. But when does the natural world ever use dollars and cents, and how does work interfere with those who surround me every day? I just wanted to hike a trail in the foothills on the weekend. Did I do that? No. Instead, I watched more television—documentaries, but still television. I developed skin eczema and migraines. My Houseplants, who I love, suffered and were not doing well. I needed more sleep while also being unable to sleep through the night. And something that truly upset me: I stopped offering moments of meditation, gratitude, and thanks for all in my life and all that I am connected to after practicing yoga in the morning. I rationalized that it took too much time. Expressing focused, present gratitude was taking too much time—what an awful mindset to attempt to justify.

I became lost while thinking I was on my path to do what I felt was right for the natural world. In severing myself, I became disoriented to my purpose, as if I had no relevance, cut off from the oxygen that inspires me to do the work I do. I was eradicating my true self while trying to save the exact same thing I believed I was connected to. The result was disconnection from myself and from all that I am.

It is so easy to slip away. I think being tethered so closely to modern life can make us more resistant to Nature. The question is one of balance. Perhaps those of us who have experience in natural areas feel them more profoundly because of our comfort with our sensitivities to the energy, experiences, and emotions of others. I sense it has something to do with our level of empathy and our own knowing of self through self.

When a pull happens, and Nature reaches out, even to those who may be limited in emotional intelligence or empathic ability, inspiration still comes. All individual lives have strengths and weaknesses. Something beneath our modern world hums lower under the "fluff and frill," beneath the facades and fears. Our powerful connecting force—empathy.

Just as inspiration is contagious, negativity spreads like a cancer. By associating our connections to life in Nature as independent from who we are and from our support system, we are, in essence, declaring these lives irrelevant to our own. Impossible as that is, given that we share this planet and are Nature, too. Regardless of our beliefs and the importance they hold in our lives, Nature exists as a community of life on this planet, sacred in their own right. In those moments when we are faced with circumstances that shake us to our core, we may question much. The process of finding our greatness, our inner guidance, and our intelligence—and holding it within our sacred space—starts and ends within us.

Slaying the Messenger

Not fully understanding where it would go, I did feel as though I was closer to understanding my path, although I often had times of feeling my way through the brush. It seemed to be an evolution for me, one with many steps and transitions; trying and failing and trying again. All required and with reason. Then I began to feel stuck in other areas of my life.

Feeling stuck limited my sense of freedom greatly, and I began ignoring myself, ignoring what I sensed, and ignoring the direction and guidance. In the process, I began shutting down and closing lines with life in Nature. My path, my purpose, became more thought than the familiar pull from within. It was unnerving, and I did this unknowingly, even when I had the best intentions. I still had a deep appreciation, care, and love for Nature and life in Nature. But I remained unaware of the two parts that completed my sense of self. One part was mine—my active role and participation. The other was not a role to be filled by someone or something else I knew, was with at the time, or had yet to find. Not of children others wanted for me, or the shame expressed toward me as I felt compelled to explain myself away, while feeling depleted from leaking energy from within. I learned that having children for the wrong reasons could be most selfish and not the path of one who perhaps wanted to add to life differently. Different does not mean wrong.

This complete self included myself with Nature, facing one another in a way I had never considered. Within me, this was where I was not providing a suitable and safe place for the existence of my other completing half—life in Nature. These voids left me like a sponge, absorbing so much negativity around me. All of us can feel when something is lost — no voice, and without greatness.

Paths are best understood by those on them. Branches hang very low along the way, and turns are filled with beautiful growth and blinding bright sunlight. I believe purpose provides momentum within us, within our cells, and is fueled by relationships with life in Nature. When we have personal relationships with life in Nature, we are supported and guided in what we believe and hold dear, at times flooding with inspiration. Without these relationships, we become reactive to the other demands in our lives; very functional, even transactional, and very limited in depth. We can become unbalanced and without grounding. It becomes difficult to keep up with the demands that begin to push against the energy and support we have within. Slowly, we turn our backs away from Nature. What we need most, we think we need least.

The lives throughout Nature create a support system, receiving and sending information. Perhaps we could learn a great deal about ourselves in the presence of their greatness. It is so simple among such complexity of intelligence and emotion. Support for our greatness—our inner intelligence and our purpose—finds its source within our treatment, understanding, and acceptance of the purpose in the lives of Nature. Our motivation comes from within but must have a place in the world to be visible beyond our being. Known in our empathy, are we up to the task to reciprocate? Perhaps we would learn if we stopped for a moment, set down our knee-jerk reactions to slay the messenger, and simply...listened.

Power Distorted, No Power Within

It's easy to focus on those who bully, hurt, and yell—sometimes at the top of their lungs—all that is the opposite of real confidence and glaringly apparent insecurity within. What's really happening is that we miss Nature and all that this means. And it means a great deal: our perception of power is distorted; our idea, notion, and goal of what power is distorts us and our vision of life in Nature.

Our inner health and outer support system—the immediate world in which we live—are connected through our experiences and begin to deteriorate based on cruel, hateful, and hurtful intentions, motivations, and actions directed outward from ourselves. We direct this toward life around us and our environments, including our homes. After all, actions are motive and intention in energy form. They become alive in our lives around us. And if they are directed toward Nature, they are directed toward ourselves and toward other people, too. With this distorted sense of power, we wobble, and there is no footing for supportive kindness and trusted respect to create the lasting greatness that feeds the support of our purpose.

We have become people who are quick to anger and willing to share great negativity. We doubt ourselves when we do not understand our world. We create this by cutting through the messenger relaying understanding. We attack others or falsely blame them, as if someone needs to step up and take ownership for another's chosen awfulness. For some reason, attacking—along with the insecurity behind those attacking emotions—has become acceptable as a form of strength. Kindness, love, and acceptance have been pushed aside for too long or fly under the radar, rarely noticed. Maybe it is because these qualities are so much more natural and therefore need no attention beyond their state of being. Authentic kindness is void of insecurity. We are uninspired, and we continue to attempt to drain inspiration from others.

Still, many of us search for hope and support and seek it along the way for others, too. True confidence and strength need no spotlight. We need to look for inspiration. If we are left uninspired, if inspiration is hard to find, the void begins to be overrun with discouragement. Discouragement can breed resentment and, if not addressed, it takes over and spreads uncontrollably, distorting everything in its path. Perhaps the messenger of our greatness has been discounted by us all this time.

Trust is our core intelligence—our trustworthiness, that is. We have lost how to honor trust and extend trust. Understanding the power Nature has on our lives, as a source of inspiration alive and flourishing with us in the present, could become our tool for understanding, our support for calm, and the healthy inner world that our minds and bodies require. When considering power in its truest form, turn to trust. Not spoken trust, but authenticated trust. Trust is the most powerful force in any and all relationships. Without trust, there is no stable ground from which to flourish.

Our relationship with the lives throughout Nature is plagued with dysfunction and toxicity. Nature is always trying to return to a state of balance, yet we seem to fight this at every step. We seek gratification outside ourselves when it is our role to help provide it from within, including through our connection to Nature. Yet we have broken that trust with life throughout Nature. Attempting to complete one or both halves of ourselves through means entirely outside of us—destructive, altering, or distorting our senses and perception of what we need—creates imbalance when done without Nature. Trust matters most in all relationships, including our relationship with Nature and our inner relationship with ourselves.

Fueling our inner intelligence, negativity does not create positivity—not within our bodies and not out in the world beyond them. Negativity spreads more negativity like a flood, slowly engulfing anything and anyone in its path. We have not been taught that relationships require mutually supportive balance and a profound respect for the rationale behind our actions. When this is not honored, we are the ones who suffer greatly—physically, mentally, emotionally, and perhaps spiritually. There are times when we must let go of what we think we need to be doing to allow for what it is we truly need to be doing. We will continue to struggle without the natural world at our back and in our hearts.

Leaving Hope Behind in a Meadow

How many times have we acted in ways based on what we *think* is needed instead of asking what is needed? We struggle to do this with one another. Often, we don't take time to do this with *ourselves*—in those quiet, calm, peaceful moments. Chances are, we never stop to feel what we need while sitting peacefully and quietly with an individual in Nature. But could we trust ourselves enough to learn what it is that we need instead of steadfastly insisting we already know? If it is absurd to simply consider, then maybe that is why our meadow is without hope. It is in this unwillingness that we lose touch with our inner intelligence, our greatness, and our purpose, when anger, control, and dominance flood in places that only function in balance, support, and cohesion.

All relationships in our lives revolve around trust, including those with Nature. The ways void of trust—such as manipulation, deceit, displaced anger, unresolved issues stewed in hate, and inhumanity—scream loudest in Nature because here, manipulation used by human motive and words have no home. It is returned to us slowly. Trust reigns supreme when it is our true standard. Our integrity is heard in Nature.

We are more alike with life in Nature within our bodies, brains, experiences, and hearts than we are not. If experiences bring people together, would it not apply to our experiences with a life or lives in Nature when respectful of Nature? We are designed this way, and our connection to Nature is a need we have inherently. We have lost our way and do not offer thought or consideration to leave Nature to their homes in the meadows and mountains, and instead embrace Nature in our homes—yards, porches, balconies, streets, neighborhoods, and communities. Where there is hope, anything is possible, and our behavior with the life around us, including life in Nature, impacts our attitudes, and our attitudes impact our behavior. This cycle, or circle, can be of our making. There is nothing great in anger, control, and mistreatment. We are responsible for limiting our positive experiences. We can and do drown in negativity until we become it. Thinking too much can make us sick. Experiencing and feeling kind emotion is healing.

When our needs are not met, we struggle. This is true among human and nonhuman lives. We are without grounding strength. Life will always toss us into times of uncertainty, and we feel lost. Hold on to knowing our greatness. When moments that touch softly and inspire greatly are embraced, reciprocated, and directed back out to another life in Nature—to Tree and sky, to another person—change happens.

In our struggle to add value to our lives in fundamental ways, we leave all thoughts, care, and concern for life in Nature behind. The lives of Nature around our home would begin to know us well and we them, if we cared to offer them the trusting space they need. We are more alike than not, and our value of self would benefit from stepping into this light. Perhaps our emotions have now become so inferior that false hopes circling around us and within us are preventing us from being all that we are and radiating this out into the world.

We know within, in our own voice, our own light. Nature draws closest to what is real.

Let's not hold ourselves back from our dreams. Want good for everyone, holding the knowledge that each one of us makes our own place in the world—of and for good. Honor inner intelligence. Place trust in this and those who are of our trust's worth. We need our uniquely individual quiet, calm, respectful relationships with individuals and all life, near and far, through the breeze, and Earth, and Trees—Nature.

Sister Rabbit

I moved slowly one morning, sensing that doing so was the right thing to do. I had been leaving Rabbit alone, but I always seemed to be drawn to where she was resting unknowingly and startled her. She was the sibling of very small Rabbit, Lemon. I did it again that morning, but she or he hopped slowly to their favorite low-growing Tree and rested in the shade.

After we crossed paths, Rabbit reminded me to respect space, boundaries, and keep a trustworthy distance as she or he has defined. That is what Nature needs. This is what Rabbit needs. So this encounter was a good reminder.

Rabbit, along with other Birds and Squirrels, has been a significant individual in my life. They are all significant teachers. I hope to return the kindness to them, too.

We do not need to leave our hope behind in the meadow that has inspired us so. We can breathe this in each day and live life with inspired hope flowing back out from within—our inner landscape.

Once we leave all the beauty we experience in the meadow, we can retain that hope offered to us. Our greatness from within, supporting our purpose, affects all areas of our lives because it affects all areas within us. Purpose creates good for the future and hope for the future. Our greatness and purpose come from clear inner intelligence—supported by Nature. Nature within us and around us. The question to ask within our own hearts: what would we become or lose if we learned more of this trust?

Part Two

Looking in the
Mirror of Nature

Eight
Nature Communicates

Offering Respect to Nature

Nature communicates—and for the most part, we have been ignoring what our senses have been experiencing. Without words, we can offer attention to these relationships in a way that supports them and in a way they deserve—where we consider what they need. And perhaps they need to be left alone, in peace.

We must offer respect to the lives that *are* Nature. In doing so, the gentle voice of Nature flows back to us through our senses, and if we are wise to listen, we have much to learn.

We have an inner need for a special, personal connection with life throughout Nature around us, our homes, and in our communities, where we work, shop, play, and live. These are not far-off places; they are our everyday environments—places we no longer escape to find an escape. Places, perhaps, we need to be trusted; possibly in ways we have not been so trustworthy. In our support and acceptance of these lives, we are supported. Our mental and physical health is supported. Our goals and who we are become supported.

Nature heals because our inner Nature heals. We are different in wonderful ways that connect us as the same.

This is wonderful. The avenues of communication of interconnectedness might be different, but we are not at a total loss. What we sense, appreciate, and extend on trust does not make the reciprocated expression any less profound.

Conversations

When we allow ourselves the time to slow down, calm our minds, refrain from speaking—and from assuming—and open to what it is that life in Nature might be offering to us, we offer the space life in Nature needs, too. Gently, required space is priority for their own, unique voice communicated in their own, unique way. We begin to learn to "hear" these lives through feeling.

Our senses respond in a natural, innate way. Sometimes, there is an overall sense of the moment as another goes about their day. *Calmness, content, hunger, curiosity, peace, play, joy.* At times, that subtle communication is for us directly. *I was hoping to understand Chickadee ways of finding food, but instead, out of no thought of my own in that moment, I felt determination to never give up on finding what makes me happiest, and that it will be found in joy.* Nature communicates. We learn about life around us. We can witness their wordless conversations among one another, then something incredible happens: We learn about ourselves.

Robins communicate among Robins. Redwing Blackbirds communicate among Robins, too, along with Song Sparrows, Crows, Goldfinches, Bluebirds, and all other Birds. I have watched Hummingbirds interacting with Honeybees and Bumblebees and Coyotes. I have watched Belted Kingfishers communicate with Flickers. I have been next to Squirrels communicating with Hawks, Owls, as well as Goldfinches, and Rabbits. I have watched their behaviors adapt and adjust in advance of weather changes and systems. I have learned more about the seasons from their activity and sensory responses. I have seen Trees respond to the presence of wildlife visiting around them. When basic needs are met, I have watched a considerable amount of play and joy, peace, and relaxation. Life in Nature communicates with one another without words ever uttered.

A knowing transfers from the incoming breezes. These conversations are communicated in ways I do not understand fully, but they seem to flow with information full of intelligence. They do have a common language among themselves and understand other forms of communication they have learned from others.

I have had Squirrels understand sounds I was making to Willow, while I have misinterpreted those same sounds. For Squirrels, one sound was a warning call, and others followed suit. Aware of this, I adjusted quickly. Once relieved, quiet returned to play and a meal.

I am convinced these more silent conversations are attempted with us, at times, in certain moments; in long, silent stares of intuition initiated by Nature, and with trusted individuals. It is not ours to decipher fully, but it is up to us to feel them so we might know better. We are missing out on many conversations, and our minds appear to interpret them first, given our current physical and mental health struggles. Our minds, our consciousness, are starved of these other forms of connection.

This is not to say that we *must* understand what is being expressed by other nonhuman life around us. I do not think that is even possible given the diversity and means of expression, sensory skills, and strength. For instance, Dogs communicate in such subtle ways with one another through body stance and shift, ear movements, most of which are lost on us. Yet, they have learned to understand our language a good deal and work with us in our relationships.

We all have strengths and weaknesses, including life in Nature. What is glaringly apparent is for us to acknowledge that the lack of human vocal cords does not imply inferiority in expression and communication abilities or intelligence and emotion. We can, however, learn to strengthen our crippled and dulled sensing, our suppressed emotions, and our struggle to have our own voice revitalized with life in Nature. If we are willing to listen, we would know these lives differently and that each individual is very different. We might begin to understand our strengths in ourselves.

Interconnectedness never stops. It is ongoing. Experiences create a *sense*, triggering an emotion, which formulates a thought, even if hours later, and delivers our experiences within us in a very real way. So, our ability to remain open must be incredibly positive. Turning positive into something negative disrupts the functions in us in ways that do not serve us well, physically and otherwise. Both will impact our brains and our bodies, as well as our relationships with Nature. What is fascinating is that interconnectedness is happening without our knowledge, too, and that consciousness begins to step into the spotlight. Electrons in blood appear to work together as if they were part of a larger whole. We *are* the whole—and we can follow the tiniest leads within and around us to be more mindful of this.

Conversations are going on all around us. It is possible that in our moments with Nature, these lives are trying to connect with us, too, in their way and point of view that works so well for them. Perhaps, in these relationships, we can find comfort and understanding, direction and guidance. I have certainly found this, and I know other people have, too—not in profound flashes in far-off places, but calmly at home and in the areas we live, because this is where we are known. These moments are amazing, life-changing, and profound. Many are already reinforcing our common brotherhood and sisterhood on this Earth, quietly and respectfully.

Unknown and All Felt

Around 7:45 a.m. one morning, I ventured out back and took the feeder out of the night storage can along with the feeder cup. They both had food from the evening prior. I turned to walk them over to their stands, and Birds started flying around me and over my head and straight to the feeder.

I was right next to the patio and water basin. There were about ten individuals around me, talking softly, flying a foot between my body and the feeder, landing on the dome of the feeder, trying to land on the cup...eating, and eating. I moved to walk, and they flew off as I moved respectfully and slowly over to the feeding station to place each feeder on its hook.

One Goldfinch flew back onto the feeder... And then another, and another. All of them came back while I was holding the feeder still. They were cracking seeds, chatting softly, landing on my arm and hand. I moved slowly, and they landed up next to me on the wood fence. I placed the feeders into position, and after I went back inside, they did not come back to eat but flew up into the big Trees off to the right, about a hundred yards away.

They did not return in the group or in the near future as I peered periodically from inside through a window. I went back out after a few minutes and held the feeders, but they did not come back. I tried again the next morning. They did not do it again. At this location, it happened once in about three years. Regardless of why, the moment's expressions were powerful.

Information is communicated in other ways above and beyond verbal or vocal. Nonverbal and nonlocal information—connections regardless of distance and understanding beyond what is apparent in a present state and place—can be the loudest for some wired in these strengths, if allowed to hear, sense, know, and understand.

We Create, We Radiate, We Are

How we think about and interact with Nature impacts these lives. This impact radiating from who *we* are, reverberating back to them, continues as we have been doing in devastating ways. Let's consider that by doing so, in the process, we are doing the same to ourselves.

We do not exist in a vacuum. In these moments when our minds are calmed, we can focus on relaxing the tenseness trapped within us from our worries about our days that are held within our muscles and minds, and direct our focus to being present within our bodies. When we are responsible for *where* and *who* we are, we place trust in life around us. Every single effect within our world, upon our Earth, has a cause—an original starting point. In learning to trust in ourselves, we start the process of being trustworthy.

Learning to understand another in the natural world beyond our capabilities creates the stage to learn more about ourselves. We are a part of a larger community, and acknowledging this world allows us to evolve with and not be cut off from Nature. We learn to begin to evolve in healthy relationships with life in Nature. With honesty in ourselves—in that self-trust—we are creating our fertile ground from within to life around us; interconnectedness. What we learn, we adopt and pass on within our human relationships. The law of cause and effect is universal; every single action in the universe produces a reaction.

If we are to learn from Nature what it means to not only demonstrate or show qualities such as acceptance, compassion, care, consideration, kindness, but also be more of these ways from our within, our interactions with life in Nature will allow for positive relationships to root and expand. We will welcome what we sense in return as joy, gratitude, ease of mind, play and fun, and appreciation. Our cells—our inner Nature; our physical level; this cellular level—as well as within our psychological mental health realms of who we are as a person, will respond accordingly.

Our brain functions and sense-emotion-thought relationship, traveling through us from our experiences, emotional intelligence, and mindful understandings, impacts our cells and our bodies. Our experiences stimulate a sense (or senses) resulting in emotion—often in a flash so fast, which triggers our intellect and thoughts, which then kick our bodies into action. It is during this process that we can first calm the mind and brain activity and allow another to have their own experience. Now, we are *listening* to Nature. Like any skill, we can learn to do so more and more to begin to know and understand unique individuals.

Life in Nature has been communicating all along. Perhaps Nature knows their communication has fallen on deaf ears for far too long, for we have not been collectively willing or ready to listen. Through our interconnectedness, what we often believe through our thinking brain is not always what we truly need with individual lives in the natural world. Through our senses is what the other parts of self seek. Here, we evolve by including ways that have since been lost.

Above Our Heads...and Below Our Feet

Plants do communicate in their own way underground, through the breeze, with the mist, and water droplets. Plants are unique in that they do not have faces like Animals or Insects or Marine life, nor do they move location. Not that having a face has softened our treatment of another being. Plants do move positions to embrace sunshine, as many who care for Plants can attest.

Plants do not scurry about or run away, take flight, or jump in the air. Perhaps because of this, we could see Tree on our property, outside our window, in our neighborhood, or on our way to work, who is hundreds of years old, without any impression made upon us. Most of us do not notice these lives all around us.

Many gardeners and those who care and love Plants in their homes would disagree that Plants do not offer much. Providing for so much life, both human and nonhuman, affords them high honors. Plants are more than producers of food or available for our consumption as nourishment, building materials, or otherwise. Plants are much more than this. And that is where their highest honor flourishes.

Our ways of viewing these lives have failed them and us terribly. These wise elders do share their wisdom through Earth. Maybe if we knew of these lives in terms of who they are, then we might do right by them, and in turn, heal our spirits and our scorched landscapes before it is too late, and our elder Trees, along with all of their wisdom that younger Trees need to survive, are gone.

Plants have memory in a way and through systems that work for them. They know when to bloom based on memory of experience. They adapt to their surroundings to possibly feel better so they might survive. They respond to aggression and to kindness.

There have been studies demonstrating that Plants seem to know when someone who has hurt them is approaching again and will emit a warning to other Plants through chemicals and energy waves. Perhaps they are "red flagging" some of us, and maybe these signals are sensed by local Insects and Mammals, as well. That is their expression of fear. Plants cannot run, but they do remember and learn. They have an awareness.

From my experience, all life in Nature seems to be aware and remembers. I have moved a lot and have witnessed and documented how my treatment of life around me has opened the doors for me to new homes, often in locations thousands of miles away. This knowing of our intent, and maybe who we are, spreads. Perhaps our actions follow us. How do we want to be remembered? How are we known in this larger community of life? How do these relationships circle back to us?

If relationships grow from our communication, along with our connections that we have formed with ourselves and other people, then this can have a significant impact on us and alter who we are, along with our health and happiness. Why would the same not be true about our relationships with life in Nature?

Kingfisher Helping Flicker

Shortly after I filled and set out feeders, I sat on the deck during an early Autumn morning with Willow wrapped close to me on my lap and in my jacket to stay warm. We sat at the bistro table behind our home.

I heard Flicker, who frequently visited our yard. This was one of two whom I knew—a male and a female, who were together often.

I looked up at the tall Trees that Squirrels rely on, and Flicker, the female, caught my eye to the right. I watched her fly off north with the male Cooper's Hawk.

He was in pursuit of her. I watched her take a sharp dive to the right and drop, and the

Cooper was thrown a little by this move, but he also dropped in the same direction.

Then, a third individual appeared in the sky: our Belted Kingfisher rattled on in. Rattling and rattling, I watched Kingfisher chase behind Coopers Hawk, gaining rapidly. And then there was a shift in the sky and with flight paths, and Kingfisher allowed Hawk to chase him or her (and without much worry, it seemed). Flicker had diverted and was safe in Trees at this point. Cooper's Hawk gave up, and Kingfisher flew south and slightly west a little, still rattling.

Time and time again, I have witnessed Birds of the same species or of other species assist one another, such as Jays and House Finches protecting nests together, defending against attacks with alarm calls, dives, and circling maneuvers from determined Hawks—like Goldfinch, Redwing Blackbirds, Hummingbirds, and others. Perhaps it is to let everyone know, while chirping, almost causing too much noise and swirling confusion for Hawk, while signaling that they are "here."

I have been two feet from different Hawks on fences in different locations. They seem to lock in on smaller Birds, in moments hiding in Grasses. I became invisible. It seems like they have an ability to focus and see in ways that filter out everything else. It is as if when there is too much noise and energy from activity around them before this happens, perhaps the ability to focus to allow for their gifted senses is thwarted.

Scrub Jays give warning calls to areas to let everyone know of a potential threat, like so many in the natural world do for others. I have watched one individual help and protect another while risking their own life in the process. A few moments later, as I stood to return inside, I saw Flicker, who was safe in Trees, fly off in the same direction as Kingfisher. Perhaps their relationship was deeper than I am supposed to know. Regardless, I witnessed one of trust.

Messages in the Nonlocal

Nonlocality seems to be occurring in our everyday lives. It seems that if it has been theorized to happen across light years, where a happening in one location can be sensed in another, why not in moments when we are cooking our food, walking our sweet Dog, driving home (and our Dog knows), meeting with people, or sitting outside in silence with our gardens among Trees?

As we remain open to possibilities, we begin to know what it means to have neglected one of life's very basic needs. For me, I have learned, witnessed, felt, and believe this collective consciousness, cause-and-effect reality, and the nonlocal information and unexplained awareness and knowing—is the essence of interconnectedness and is as basic as the other defined needs for which all living beings require. For us, to remain aware of our interconnectedness is what matters.

There is nothing unusual we need to do to embrace this into our conscious life and world. What does matter is to remain mindfully aware when we are with life out of our doors because this world is very much present. Because we are part of the larger equation along with the rest of life on Earth, our internal life and struggles are felt. Therefore, our struggles become their struggle, and their success and balance could be ours, too.

Some of us are so deeply in tune with this type of sense as an alternative way of receiving and knowing information around us. For whatever reason, we have not completely lost the ability to know that seems to be present with life in Nature astutely—the nonlocal. These relationships we have with life around us provide the anchor for our senses and connect us all, through joy and love.

This sense appears to us as images and physical feelings (sensations or senses) that present themselves out of "nowhere," or create an image in the mind in the presence of another living being, and then an emotion or a thought (or both at once) occurs. From my experiences, they cannot be manufactured, imagined, or created. They are gentle at times and nagging at others. Outside of us locking in to those who can know.

Nature is highly in touch with nonlocal information that leads to heightened and extreme knowing and intelligence. Life in Nature is very flexible and fluent in this way of understanding and knowing the world and those around them, including humans, as well as weather systems and events.

To be considerate of other ways of living on this planet and how these natural individuals, like us, have personalities, needs, and intelligence that differ from ours, and are invaluable in their own situations and existence, is to embrace a community of life on this planet. Earth is a world so unique from other planets that we are aware of, currently, and is rich in diversity of life and communities so intricately woven together in support systems. This is something we do not need to understand completely, but to stop taking for granted.

Appreciate that our planet Earth is beyond special. This planet is *our collective home*. Realize what a rare gift we have, here—Earth and our community of life throughout all natural spaces, habitats, and communities.

We can evolve as a human species if we are open to evolving with interconnectedness. But we need to *want* to change. Our health, our bodies and minds—and life around us, within us, and through us—are waiting.

Osprey, Coyote, Mallards

One spring evening, as I was taking in feeders, an Osprey flew over our yard, low—less than fifty feet in the air—and looked down as she did. I caught her eye, and she was watching me.

Later that evening, I had Willow in the front yard before bed, and I saw Coyote standing in the street about fifty yards away. She (or he) was facing me and desiring to go into the natural area next to our house.

Coyote watched closely, frozen, and leaned to go back where she came from. I remained calm and, with ease, let her do what she needed. I stopped looking so much. She relaxed and stayed centered, her eyes on me. I stopped looking altogether to offer space and trust. I sensed what was needed, so Willow and I went back into our house. I checked from inside through the window, and Coyote was gone—I do believe off into the natural area, the direction in which she was heading when we caught eyes.

The next morning, the temperature was cooler, wet with snow and rain. Birds were singing everywhere around our home and eating from the feeders. Male Mallard was in our yard, and the female was on the other side of the fence. Both were eating away. They stayed a good while. He kept close to her. Once they were ready to leave, he was struggling to push his way through the fence. She tried to help by walking up and down the fence as he followed, trying to find his way back to her. I could sense he was starting to panic a little. I went outside slowly with peace and calm in my heart to open the gate. I was almost to it (right off from our house) when she lifted and flew a few feet to the little creek about eight feet from our split rail, chicken wire-backed fencing, and he followed. I went back inside. Later, I went back out refill feeders, and they were in the creek, floating and resting. I didn't stare, but glanced, and then looked down with peace. She climbed onto our side of the creek and watched me closely. He floated and ate. They both went back to floating when I returned indoors.

I found it interesting that three Animals that are normally considered shy of people were watching me and noticing our yard. There seems to be a known that is shared among those in the natural world. They know safe places to rest and just be. This follows me now to each new location where we live. I do not understand completely, but if I stop and sense, it is very personal.

Nine
Our Sense Relationship

Song Sparrow's Request

I was taking in the feeders for the evening. I store them in locking metal cans. The Song and Fox Sparrow, along with a few others and the Juncos, will eat later than the other Birds. While walking with a feeder in hand, I heard a very loud chirping. It was so loud I stopped in my tracks and turned toward the sound, wondering who it was at this time of almost dark—and why it sounded so loud and somewhat frantic. Then, I saw him jump out onto the rocks in our yard and up onto the open-space fencing, still chirping a single chirp very loudly. Instantly, without thought or time to reason, I knew he wanted some food left out. He did not want me to put everything away. So, I turned and walked back. Once I started walking back, he stopped calling.

I placed the feeder back in its spot on the old branch we have near the ground. After I put another feeder away and went back inside, I turned and looked—just a minute later from placing the feeder cup in position—and saw him sitting on the cup eating. Once he finished and went his way for the evening, I returned and stored the cup overnight.

Life in Nature will try with us. It is we who need to sense clearly from them and act accordingly. For some, it might be second nature because what Nature is expressing matters to us. Some may try a little harder to sense the messages from information.

Interconnectedness Is Alive Within

Energy is a powerful thing. The experiences we have, how we think about them—including all of the triggering emotions, and those that linger—bind us.

Energy is radiating from all organic individuals, rippling out into the world. Energy is also created from and includes our experiences and our actions, driven by our intent, motive, thoughts, and emotions.

It is not in anyone's best interest to abuse or manipulate another (and, therefore, their energy) in any negative way. That is because the reverberation is magnified over time.

Interconnectedness is alive within and around us through our sense-emotion-thought relationship. This relationship does much for us, with us, and through us. Created from experiences, this relationship is unique to each one of us. The interconnectedness from this relationship impacts our health and happiness in both negative and positive ways. It can happen without thinking because it is the nature of being part of this world of Earth. However, awareness, or lack thereof, can create good or bad influences within us, and therefore outwardly with those around us, including the natural world.

And so it seems that our sense-emotion-thought relationship is the key to our relationships with Nature all around us, through us, within us, and back to Nature.

"The Bridge" Is Our Relationship with Nature

Our emotions are deeply connected to our body's physiology. They influence our health in both negative and positive ways. Triggered by our senses, our emotions affect each one of us in moments of interaction. Through our senses, emotions grow from these interactions, creating more thoughts. Our thoughts create chemical changes within our bodies, radiating from the brain and other areas, such as the stomach, which is home to vast amounts of neurotransmitters. This is a working relationship.

Interconnectedness comes through our experiences, activating our senses, triggering our emotions, and creating thoughts. Our experiences with the outside world—with life found throughout the natural world in our yards, neighborhoods, and communities—how we feel about someone, how we treat others, and how we are treated, directly impact our thoughts, affecting our bodies through the influence flowing from our brains.

Nature's brilliant intelligence matters in our lives more than we have been willing to embrace. We need interconnectedness with life in Nature through positive, supportive relationships. This is not Nature giving while we simply take. We need to extend and exist within healthy relationships with Nature. Life in Nature must be part of this two-way exchange.

Our experiences with life in the natural world, both individually and collectively, have a profound effect on us—on our brains, bodies, minds and consciousness, and perhaps even our spirit and personality. Positivity draws positivity, and negativity draws negativity. This is why our emotions and intelligence in this form matter greatly. Our interconnectedness with Nature occurs in its purest and most supportive form when we allow Nature to be alongside us in a reciprocated relationship. The natural world remains masterful in their intelligence and elaborate, varied senses.

An Unspoken Request for Assistance

This evening, in particular, was cold—the evening before the first truly cold autumn night.

I was placing a few walnuts on the posts for Squirrels. Big Mama Squirrel was larger than the other female Squirrels. She watched me place the walnuts from the feeder area and came running along the fence toward me. It was getting dark at this time, so this would be the last run. I felt she was pushing it a bit by being out this late. It was later than she was comfortable being away from her safe place.

She came over to me and sniffed, moving her nose among the pile of walnuts. I have seen this before. I am certain they count them; that they tally the amount to judge how many runs are needed to collect all the nuts. She did this and then ran over to the part of the fence near tall Trees that were out of our yard, far away, where she had been staying. Night was falling.

She did not take any of them but ran to that particular post that seemed to feel safer for her, as it was closer to where she would den for the night. She stood up on her back feet, though not fully on her toes, and breathed differently—like a huffing or heavier exhale. I knew immediately, much like with Willow's expressions, that she wanted the nuts moved. She waited for me, and I took the pile over to the post next to the one she was waiting on. Immediately, she went to the nuts, stuffed as many as she could into her mouth, and carried them away. I went inside, and she did come back for the remaining few on her final trip.

Note: When we were loading the truck in late summer of that same year, she rested on the stone support for our back deck and watched the entire day as our house emptied. I stood at the window looking back into her eyes, just a few feet from her. I thought to take a photograph of her, but more than anything, I wanted to remember the feeling. I do not recall the image that may have been taken. I do recall so much more.

I believe she knew what was happening—that I was leaving—and that I was feeling deep sadness and loss over leaving these beautiful lives I had come to know as individuals. We looked at one another through the glass. I think of her every day since leaving. It has been a few years now. Those final moments with her still fill my heart, my body, and my vision. She always will.

Through the Eyes of Wren

There is a little House Wren that hops and flitters about in the lower bushes when I fill the feeder or change the birdbath water. This little girl or boy looks at me and will peek from behind branches and leaves to see my eyes. I know because when I look back, he or she will hop out of sight. I have read that these fun and sweet little Birds are very shy. This little one is very curious.

This little one usually comes out and lands on the fence three feet to my left—and watches. This day, I sense that he or she wanted some seeds. So, I began placing feeders closer to the soil for those who eat from the ground.

Wrens eat seeds. We watched Robins and Robin families eat seeds at our old home in the winter when food is scarce, and through spring and summer, too. I have watched Swallows eat hulled seeds during torrential rains lasting days during remnants of hurricanes.

On this day, he or she perched two feet in front of me, watching from within Bush, still ducking down and looking under branches to see me and my eyes. I love these Birds and enjoy time with this little one. Wren remains a special force in our lives.

On a separate occasion, when I was taking the feeders down and in for the evening, this little Wren was in our yellow Plant. There were Moths in Plant feeding, and he or she may have been watching them. Little Wren watched me, but mostly continued doing what he or she was doing in the Plant. They jumped out onto the fence a few times and did this again while I was near the house. He or she seemed comfortable with me being there, too. It was dusk. The other Birds were gone for the evening or roosting in Trees. It was just me with little Wren. Special. Moments like this are special.

On another evening, I was outside refilling feeders when Chickadees and Red-breasted Nuthatches came over to watch me. The feeders still had plenty of seeds, but I did notice several of these guys going over to the cup and looking into it. The cup was empty. I noticed Red-breasted Nuthatch watching me from the fence right in front of me. I waited for him or her to move before I checked the feeder beyond the fence. A Mountain Chickadee followed me the whole time and stayed across from me in small Trees watching what I was doing. Chirping, chirping.

Sing Sing, a Mountain Chickadee, came over, too. Another Red-breasted Nuthatch wanted to get close to me but was a little nervous. I think the one who keeps coming close to me is the same one. I will call him or her Red.

They were excited that I was there. I added more seeds to the ground feeders in case House Finches crowded them out. Chickadees are very good, as are Nuthatches, at moving in quickly and grabbing what is needed, but I am seeing these two types of Birds drop seeds for others. I watched Red drop one seed for Chickadee down below, who quickly retrieved it. There were other locations for this little Bird to get seed, but it appeared he or she made a request, and Red responded by dropping one. Red grabbed a second seed and flew off to eat it.

The Mountain Chickadee that likes to get close, I will call him or her Peak.

They trust me with the food and water I offer, so I made a promise to clean the feeders and birdbaths often and thoroughly to help everyone stay healthy.

Intelligence Is Waiting

Something happened within our human development that caused great insecurity about our place in the community of life. These lives that collectively make up Nature are intelligent individuals working together perceptively, learning and growing across species and natural systems. This intelligence is sleeping within each one of us, within our bodies.

Have we even tried to experience positive intelligence when we engage with Nature in our yards, homes, neighborhoods, and communities? Could we develop that intelligent relationship with Nature and with ourselves? Can we acknowledge and actively allow life in Nature to be who they are—brilliantly perceptive individuals who experience life much like us, perhaps with deeper empathy and care?

Our senses fill us with information connecting us to life in beautiful and colorful ways. It is difficult for us to make sense of the world around us when we deny our senses and deny the sense relationships found throughout Nature. There have been countless times when I have stepped out of myself to be with life in Nature while heavy in thought, deeply focused in an area where I felt "off," only to learn that what I sensed through life in Nature offered a brighter light where I needed balance. I was misplacing my focus on a different need. Waiting quietly, Nature offered insight.

Nature is therapeutic in their own right. This is interconnectedness in action, realigning our minds, hearts, and bodies. There is a very specific sensitivity that is strengthened, or developed, through interactions with Nature—empathy. Superiority seems counterintuitive to our unified existence. Empathy is the real strength. There is a very deep, very old commonality in our individual intelligence. When honored, together we become a force—powerful, healthy, unified, and whole.

Once we learn to be aware of our presence in another's world, the kind of interconnectedness for which I speak is enlivened within us. From there, we decide how to move forward. There are ways to achieve many of our goals while honoring life in Nature. When we are in their world, we begin to understand the importance of acting in ways beyond our own wants and needs.

We can develop our senses to open additional methods of communication and understanding of ourselves, a fellow human, and our family in the natural world. We can learn to be fully present when we need to simply have a moment and just be. We can increase our intelligence and evolve in empathy as a form of high intelligence. Forced emotion is not part of the process, but true emotion is the result of what we sense. It is an intelligence not found in a book or fully understood through reading or lectures. It is a pure sensing of another through trust and respect, and it connects each one of us. We need to want to learn with Nature if we are going to learn from Nature.

Dove and Her Baby

We were walking Willow around our neighborhood, and I noticed Dove on the sidewalk beside something. She would not leave. As I left Willow safely behind, I approached slowly, respectfully, and with concern. She flew off the ground but stayed very close, in Tree. It was her baby—still growing and almost ready to leave the nest for the first time, judging by the feathers coming in. Her baby was dead, left on the sidewalk, and she did not want to leave their side. She was reluctant to leave, as I witnessed when I approached. This baby appeared to have been killed by a family Cat left outside on their own.

Our sweet indoor companion Cats become ill themselves when left in a world where they do not naturally belong. Cats, too, can become sick and return that sickness home to their humans. Companion Cats who turn feral and have more feral babies viciously attack confused family companion Cats. I have witnessed both suffering in fear and from disease in suburban and rural environments because they are not built for survival in these worlds, even when let outdoors for only a few moments, as with a former neighbor's beautiful furry feline family.

This baby Dove appeared to have been killed by someone's Cat as a form of play. The fear felt by the baby must have been absolutely horrendous—unable to fly and defenseless while his or her parent (or parents) were forced to witness helplessly, unable to intervene effectively. This adult Dove, who refused to leave, was grieving unutterably. To this day, it is difficult to revisit and feel.

I carefully moved this baby as best as I could off the sidewalk to beneath the nearby plantings and growth, so they could return to Earth and their watchful mama could sit and grieve in peace until she was ready to let go.

Animals may not cry as we do, but their sadness and grief are no less deep, no less profound. Some suffer depression so severe that they do not return.

The Voice of Nature

Our basic need for interconnectedness through our sense-emotion-thought relationship lives with life beyond our own. There can be no interconnectedness without other life found in Nature, without these relationships. Interconnectedness includes all life communities. Nature is crucial.

We need these individuals to sustain us, but we need them well beyond self-preservation and consumption, and in a way that is much more balanced than the relationship between humans and Nature has been in recent times—a more balanced relationship, void of words yet holding the power behind them. Life in Nature and their communities are, in their own rightful ways, a source of integrity that extends from a place of kindness for which we are so capable. One that is no longer one-sided. It is here, at this mindful and emotional meeting place, that we might begin to know Nature. In meeting, we begin to know unconditional support that honors our common sense and goodness within—the voice of Nature.

These relationships must be approached with intent that is honest and clear if we are to change, if improving our life in some or multiple ways matters to us. It takes confidence and acceptance to allow others to be who they are in our presence without fear, threat, or violation of self-worth. In interconnectedness, there is no room for judgment.

With life in Nature, we can be all that we are, truly, without facades, and open our senses to experience *real* emotion.

Emotion is not about crying or overreacting—common thoughts and, at times, ill-advised putdowns when someone is considered "emotional." In fact, there is so much more linked to our senses that is nothing short of intelligence at work as we experience life with Nature emotionally. We are not alone within, but we need to create a welcome mat around us for those beyond us. What comes back to us from beyond often mirrors and aligns with what is within us, even in ways that help us rebalance, grow, and evolve as individuals. We must learn in life, of life, from life, personally and deeply.

The human brain changes its thoughts into thousands of chemicals every second. So, we have to be in a good mental place. If we approach these relationships with anger, hate, or any other toxic emotion, interconnectedness will not flow readily to us. These emotions require more of our thoughts, more of our brain's processes, and more space. Otherwise, the cycle will only continue toxicity from beyond us to within us and then back out beyond us again. Anger and hate hurt us physically and mentally and pollute our brains with toxic matter, literally. It is our choice. The process of happiness is something we each owe to ourselves. We can be with life in Nature while in pain. That is very different than bringing pain to Nature.

Hummingbird Has Spoken

Willow and I were in the backyard. We did not have a feeder for Hummingbirds at the time. I put one up after this experience, and I have had one ever since.

Earlier that morning, I was hand-watering new Plants. The small sprinkler was running briefly to help in the backyard, and I watched from the deck as Hummingbird—a little green Hummingbird, who I believe was female—watched the water, flew through it, and made what looked like attempts to drink from the base of the water stream.

I had noticed Hummingbirds when I would put out the other seed feeders from their evening storage. They would watch and inspect the seeds.

That same day, while Hummingbird was flying through the sprinkler, she came over to me and stopped about two feet in front of my face. I was standing in the yard and stood still, calm. She then gradually lowered herself in a very slow and deliberate movement and moved another foot closer to me, turning slightly to see my left hand. I turned my palm out to show her an empty hand and moved it closer to her as she moved closer to it. She then did the same thing with my right hand. That hand held what I used to clean up after Willow. She inspected it as I showed it to her. She was very close to my hand. Once she inspected both hands, she flew up to my face in the same slow movement and, with clear intention, was moving around in the air before flying off to the natural area right next to our home. She was attempting to explain something, and now she sat and waited to see how I would respond.

I cannot describe the feeling. It was somewhat similar to the feelings with Bumblebees. Like the energy created by the air ripples of her wings could be felt so closely that I could almost see them, and the pull of her energy wrapped into mine and was then taken with her in a wonderful way as she flew off to wait. She manipulated my energy. It brought me to tears. I put my hand to my chest and gasped. I smiled and stood there for a few moments, and sensed all that I could—an unspoken "thank you." And then it came—I knew what she wanted in images in my mind, void of thought.

Hummingbird communicated through senses to emotion, leading to understanding through thought—our interconnectedness of experience. She wanted to know what was in my hands, as she had watched me put out seeds for other Birds. She was showing me in the same way as Willow. She was wondering if I could offer something to her through my hands, too. She spoke this in her own way—in the way of Nature.

I put out a feeder for her on that day and continue to offer Hummingbirds food wherever we live ever since.

The Leap from Experience to Interconnectedness

Our relationships with Nature are a reflection of our relationship with ourselves. Our struggles and our joys are mirrored in some way. There is a mirroring of what is perceived to be true of the self. These perceptions, bad and good, permeate relationships beyond the self, including relationships with Nature and with other people.

Developing a healthy relationship with ourselves through the positive support and understanding found in Nature can mature into a deep sense of personal satisfaction. Letting go of preconceived ideas of how life should be lived and allowing life to live as needed changes something within us. Life beyond humans and within Nature opens interconnectedness. We learn more about ourselves and our present state, and our present needs, by being with life in Nature.

Remaining open to what we experience, and not to what our mind is telling us we need or how we need to change or who we need to be, allows us to offer the same openness to life in Nature. We let go of our need to control in harmful ways when we stop doing the same to ourselves. We begin to radiate the change. We see ourselves in a new light by allowing Nature to be seen for who and all they really are, completely.

Our world within our minds and bodies thrives on these relationships when they are positive and kind, opening a doorway into relationships with Nature. By continually creating an environment in our hearts and minds that lets Nature teach us instead of making Nature become what we want, these relationships with Nature solidify and grow. Together, we think, operate, and become one with Nature in daily life.

When we begin to see Nature as an extension of ourselves and part of our world, a level of reciprocated support and trust develops. Love grows undivided. We are more likely to protect and support that which we love—those whom we love. Nature is already *a mirror of self*, an equal worthy of our care. We need to seek to reestablish a long tradition of broken trust with these lives throughout Nature and prove, without words, that we are trustworthy.

Bumblebee's Lavender

I watched Bumblebee as I was in the yard working around Flowers who were very new and not well established. This little Bumblebee appeared tired or struggling, possibly frantic. These Flowers were not helping with what Bumblebee needed. I watched as this individual Bumblebee fell to the ground several times. I guided sweet Bumblebee onto a leaf, and as I walked across the yard to Lavender, Bumblebee turned and faced the same direction. This Lavender is loved by so many different Bees and was full of buzzing activity that day.

Once there, Bumblebee walked onto Lavender and quickly started to do what was needed. I left them all alone and went back to where I was sitting.

After a good while, still watching Bumblebee from a distance, and enjoying Lavender with Honeybees, I went back over. As I reached Lavender to watch a little closer, this now content individual flew up into my face, around my eyes, back and forth in several passes, then around my head. This was done with great energy and life as they returned to resume what they were doing with Lavender among many Honeybees.

Bumblebee was doing much better now, and I felt a wonderful energy coming from the buzzing of their wings—as if they were connecting with my own energy, making us one. We were communicating. I am certain Bumblebee was expressing something in their unique way. I sensed nothing short of gratitude...a moment engulfed in great love. This would not be the last time I was privy to one Bee or one Animal expressing gratitude.

The power of the wingbeats of Bumblebees and Honeybees, when close to our own energy field—or perhaps any natural being radiating energy—feels like a form of communication. There is an attempt to convey information in these moments. Sometimes, it is gentle support, perhaps strengthening interconnectedness. Perhaps all Bees, and even others, convey information through the senses when energy fields cross.

I can feel it all over again, even in this reflection of the moment. Through all the senses stimulated by this communication, I understood gratitude. Bumblebee communicated in that moment, interconnectedly. The feeling was the same as genuine love.

Ten

Interconnectedness Unspoken

Our Intent and Actions Matter

Interconnectedness is an ever-present reality that exists independently of our personal beliefs and decisions. Though often unspoken, interconnectedness is shaped by our words, our actions, our intentions, and what we do—or do not—carry within our hearts.

We are not separate from interconnectedness, nor are we immune to its effects. Believing that our way alone is the right way dulls the support that Nature offers. Much of the loss we feel may arise from how we live, and from the ways we have chosen to relate to the life around us throughout the natural world.

Interconnectedness is always present among all forms of life, including human life. It is up to us whether we choose to embrace this natural need. Like the air we breathe, our need for interconnectedness cannot be removed—it is impossible. Like the air around us, through a breeze or a cooling wind, interconnectedness is significant to our hearts, minds, bodies, and the attitudes and decisions that shape our lives.

The energy generated by our thoughts and intentions—flowing through our experiences, our cells, our blood, our organs, and our bodies, just as it does in every life in Nature—can either strengthen our inner foundation or slowly erode it. This process of reciprocity is always at work, even when our own struggles prevent us from fully receiving its benefits.

We are *impacted* by the way we approach our relationships with Nature. As individuals, the place we give these relationships within our personal landscapes can be either healthy and supportive or toxic and harmful. Interconnectedness speaks through the unspoken voice of Nature. Our intentions matter. Our actions matter.

Witness to Yellow-headed Blackbird

Yellow-headed Blackbirds are not common in my area, and I believe this individual may have been passing through. I had to look him up based on his distinctive physical characteristics. This was my first and only time seeing him. I was very fortunate to notice him as soon as he arrived in our very small yard adjacent to a protected wetland. I watched him stay all day, checking in on him quite often, and observed carefully and respectfully from a window. He was interested in resting—sleeping, eating a lot, drinking often, hiding, and resting.

At that time, our yard was little more than barren clay with scattered non-native and invasive plants growing in the dry heat of a parched summer. It was a prime environment well suited for invasive species to overtake the native Flowers, fruit- and seed-bearing shrubs. Even in such a small space, I fought with these Plants through weeding by hand, but left many as cover for Birds until we made a suitable resting place for these lives seeking shelter. We occupy this land at their expense—something I hold in my awareness as we respect their needs for distance and safety.

I placed a small feeder and a tiny birdbath to support both those migrating and those who chose the surrounding area as their year-round residence. Water is vital to all life, and our birdbath is used not only by Birds, but also by Honeybees and other Insects. On that day, I was in the right place at the right time when Yellow-headed Blackbird was ushered here to this very small space in our care.

One of our male Red-winged Blackbirds, who had come to know and trust us and our yard, flew with him and guided him to our welcoming space. I watched the exchange of understanding from Red-winged Blackbird to Yellow-headed Blackbird before Red left and Yellow stayed.

As I observed from our bedroom window, I could sense nothing but someone asking another for assistance for food and water. Perhaps they were both off-course or traveling from a greater distance. What I witnessed felt like a simple act of kindness: one life guiding another to a place protected and trusted by Birds in the area. This little space—our yard—offered clean water and a feeder filled with hulled Sunflower seeds. One individual was in need, and another stopped all they were doing during mating season to help because he knew how. Together, along with the rest of the natural world, they matter—and they matter to one another without a vocalized word spoken among them.

Nonlocal Information

When we allow the negativity of others into our lives, we are altered physically, mentally, and emotionally. Knowledge and mind can collide and contradict one another: Language, though capable of healing and connection, has also evolved into a tool that can be used to deceive and to wound. People often hide behind words or use them flippantly. Life in Nature is more aware of the presence—or absence—of a genuine relationship, sensing motive and intent with remarkable clarity. In this way, nonverbal communication is often more truthful than words. Behavior reveals far more than language alone.

This presents a challenge for us, as many people struggle to maintain clear and honest motives and intentions. Yet these inner states are felt both by the life around us and within us. At our core, we are always seeking unity with the world around us. It is difficult to stand before the lives of Nature without *our* true nature becoming visible, front and center. Nature is wiser in this regard. Nature seeks balance and calm, even for us—the ones who have caused, and continue to cause, so much damage. Life in Nature still follows the deeper principles of the natural world. The question is whether we are willing to do the same. It is here that those who are emotionally strong and emotionally intelligent begin to thrive.

With all that we are capable of becoming, we sometimes behave as though nothing of beauty or support matters. Yet consciousness extends beyond the individual self and connects us to others. Nonlocal information is consciousness. In this reciprocal relationship, interconnectedness is continually at work, perhaps in ways some people intuitively recognize.

The lives who make our yards home are more in touch with nonlocal information, perhaps in ways that are much more advanced—they are more attuned to this broader field of awareness than we realize. We have been conditioned to disregard much of what we perceive if it does not appear to benefit us. Yet Plants, Animals, and Insects possess their own forms of awareness, perception, and understanding.

Interconnectedness exists both within us and beyond us. We are not independent of it. Together, it forms a sense of wholeness and a collective consciousness. We are much more alike with life in Nature. When these relationships are welcomed and respected, a deeper harmony emerges, even if it is not always visible to the eye.

Emotions are natural. Too often, however, people decide to force them into "boxes" and label them as "erroneous." In truth, emotions are our guides. They signal where change is needed and help us move toward greater understanding. When felt, processed, and integrated, they empower us to respond responsibly and to evolve. All of us need much more acceptance and far less judgment imposed by others. Each individual has the capacity to shape who they become. Life in Nature appears to understand this. We are still learning.

Our balance is restored through a peaceful and calm mind. The relationship of balance works both ways: As we become more balanced and open, empathy becomes our center point. Through empathy, emotions guide our knowing and learning, creating an inner environment in which interconnectedness can move freely through consciousness and mind. Empathy keeps the artery open for interconnectedness to flow with life around us.

Interconnectedness both shapes the mind and is shaped by it. It flourishes through an attitude of openness and respect. To honor interconnectedness is to honor the right of life in Nature to exist as freely and fully as we wish to exist ourselves. It does not require complete understanding of another's experience, only a willingness to remain open to their reality and significance. When we recognize the worth of lives different from our own, and honor that worth, consciousness expands, and our interconnectedness with Nature begins to flow.

Crow Calling

Crow has a light gray spot on the back of her left wing. She often calls and waits for me.

One morning, Crow could see me moving through the house beyond our wall of windows. She called for me from Tree near our back door with three short calls.

She stopped calling as soon as I waved to her from the window. As she leaned forward and ducked through the branches to see me, I acknowledged her. She understood what this meant and waited patiently.

She flew down, and I tossed her a few unsalted walnuts. Over time, she appeared to have made some agreement with Squirrels who mostly left her alone.

She and Squirrels often eat together. She frequently soaks walnuts in the birdbath, eating some immediately and taking some with her when she leaves.

That past weekend, my husband was standing on the deck near the feeder area while I was on the other side of the yard watering the raised flowerbeds. I heard someone fly in and asked him who it was. He said he did not know; I do not think he had heard her. Crow has an extraordinary way of appearing and disappearing almost without notice.

Then I heard her soft vocalizations. She was near my husband and could see a few walnuts on the ground. I sensed she was attempting to establish communication. In silence, I motioned to my husband so he would notice her. He was standing too close to the walnuts, and she seemed to be kindly letting him know what she wanted.

My husband moved but remained focused on his thoughts rather than on what she might have been conveying, watching her from our small bistro table. When I walked over, she started to click at me, bowing. I immediately understood what she was asking for.

She was quiet as she watched me pick up two fresh, healthy, unsalted walnuts from the ground and showed them to her. I walked slowly to the birdbath and placed the walnuts on its edge. I tapped the rim in the familiar way I use as a signal to the life in our yard, then walked away with my back to the birdbath to give her space. Moments later, I heard the beat of her wings as she flew down, took one walnut, and departed.

My husband was a little awestruck by her sounds and her messages to us. My sense was that she first tried to communicate with him because he was nearest to the walnuts, and then turned to me—clicking and bowing—when I approached. With my husband, she appeared to be seeking common ground and understanding. With me, the message felt immediately clear, as if I could hear her without a single spoken word. It was a knowing carried through an unseen connection. She is remarkable.

Our Thinking Extends Beyond Ourselves

After I heard Owl attacking Squirrels in a nest at dawn, I wondered whether the cries might have come from Rabbit. A day later, however, I heard a similar cry from one Squirrel interacting with another and realized the sounds had indeed been made by Squirrels. That particular morning, the cries were awful and gut-wrenching. It is one of those experiences I do not like to revisit because the sensory memory of what those lives were enduring while getting attacked is almost too much to feel again.

Amid the screams, I heard the repeated sound of something heavy falling through layers of leaf-filled branches. Eventually the screams went silent, though the sound of something descending through the leaves continued for a time before falling silent as well.

At dawn, I could make out two adult Squirrels, backlit by the brightening sky, scrambling away; reluctantly looking back, over and over, from where they ran—their silhouettes perfectly outlined in the morning sunrise. They were leaving Tree.

I realize that all living beings need to eat. Many of us survive by consuming other lives, and each of us makes choices about what we will and will not take. Regardless, the lives that sustain us deserve to be honored. They become part of us, contributing to our cells, our healing, and our continued existence. This, too, is interconnectedness. When these lives are not honored, I believe we take into ourselves that disregard as well.

One late spring afternoon after work, I went into the backyard to refill the birdfeeders. Many Squirrels were in large Tree that day, and Great Horned Owl had come to visit. Judging by her larger size, I believed she was female. She rested on a lower dead branch, appearing to have spent the day there. She was sleepy, with her eyes half-closed, slowly waking to begin her evening. She was perhaps ten to twelve feet away and slightly above me. From where I stood, I sensed that she still felt safe.

This was when I fully noticed her presence. I watched her respectfully, avoiding a fixed stare. I moved quietly through the yard, checking feeders and going about my tasks without speaking. From time to time, she glanced in my direction, but increasingly she turned her attention to the world around her. After about fifteen minutes, I looked away briefly and then noticed that she had moved to the neighbor's roof, roughly twenty feet from our house and now outside our fenced yard.

She rested there while I remained about thirty feet away and within her view, careful not to make her nervous. I glanced at her occasionally, just as she occasionally looked toward me. The sounds of people, traffic, lawn mowers, and everyday activity seemed so much louder. I am often sensitive to constant noise, but in that moment, the surrounding world felt especially intrusive. Her head swiveled in response to the same disturbances.

Hawk flew in the distance and then passed overhead on a thermal—a good distance up in the sky—the only other presence that fully captured Owl's attention. She watched closely and seemed to duck as Hawk flew over, even though Hawk was hundreds of feet above our heads. Given her astounding abilities in sight and sound senses, what she experienced was different from mine. There was something between Hawk and Owl—a relationship that must be respected.

Little Boy Squirrel—a very curious Squirrel, who was around often and knew me well in the backyard—noticed me. I watched as he came to see what I was doing. I was certain he would soon be in Owl's vision, too. Given the experience from the other morning, I watched and waited in horror of what I was about to witness—I was certain he was going to be attacked by Owl. On this day, I thought and planned how I might help, but I also knew that this yard was a place of safety, where all were free to eat, drink, rest, raise their babies, teach, learn, and simply be. That balance seemed to be respected.

From where I stood, I watched Squirrel look at Owl, and then Owl look at Squirrel. There seemed to be a brief exchange to which I was not privy. Then Owl returned to scanning her surroundings, and Squirrel looked back at me before turning and returning to the yard. What had first seemed like a potential tragedy shifted into a sense of calm, contentment, and even familiarity between them. I was grateful that I had chosen to trust both individuals and allow the situation to unfold without interference.

Perhaps Owl's needs had already been met. Perhaps Squirrel sensed this. Perhaps the unspoken exchange I witnessed held a deeper understanding between them as they connected with one another. Perhaps they were both wondering what I was doing and why I was so interested in their lives. Over time, I have come to know that there is a profound respect among the lives of Nature.

When balance exists, trust and respect are present without question. Perhaps this is always the case. There is no needless torture or suffering inflicted, only the embracing of balance. This is interconnectedness held in highest reverence. We, as humans, often impose our own interpretations on the situations we observe, but what appears obvious to us may not reflect reality. Too often, human relationships and interactions stray far from the ways of Nature.

Inwardly, I thanked Owl for allowing me to join in a brief part of her evening. When I returned inside and checked again a few minutes later, she was gone and on her way into the night.

All living beings experience a wide range of emotions, though we express them in very different ways. This is true of every one of us, including the lives that make up Nature. Emotions are central to our senses and our thoughts. Through them, we begin to appreciate other intelligent lives rich with experience and expression, and to value them for who they are in their own right.

Intelligence runs deep through emotion and through our capacity to adapt and respond. Without emotions—and without life in Nature—we are starved from our need for interconnectedness that remains profoundly unmet. By staying open to knowing lives beyond our human relationships, we allow Nature's voice to be heard. Positive relationships with the natural world begin with our thoughts: how we think about ourselves and how we think about them.

From the cellular stimulations influenced by our thoughts and emotions, we connect with life in Nature through the senses we share at a very deep, substantial level. In moments when we allow these lives to express themselves as they need, our role is to remain open, supportive, and respectful, always mindful of individual boundaries, including our own. We are interconnected with one another and with all life in the natural world.

Unified Communication with the Lives of Nature

In the balance of Nature, this is not a place for our thoughts to cloud another's way of being. We can release our need to determine the "what" and the "why" and allow these moments to unfold as a form of shared communication with life in Nature. Our bodies—even our cells—respond to this connection. A need is fulfilled. This is not a place to interrupt with language that belongs only to humans. Together, in our yards, our communities, or during a quiet moment outside our offices, we discover common ground—because we live on common ground.

Viewing life throughout Nature merely as a resource to serve humanity is not respectful. When our intentions are rooted in what is truly good for these lives, and when we recognize that they matter in their own right, our respect for Nature directly supports and heals us as well.

If we had remained more deeply connected throughout our evolution—even as we made useful and life-sustaining discoveries and advancements—perhaps life would be different, easier, and more balanced.

We must begin to view our interactions with Nature as relationships. We can choose to earn their trust through the quality of our own presence. Let's stop talking, honor the community of life around us, and learn to listen. Our bodies will respond favorably. The question is whether we trust ourselves enough to offer the same kindness in return.

I am deeply thankful for these moments with individual lives in the natural world just beyond my home. Their depth is too personal to share in entirety. I wish to never regard them as anything less than sacred, profoundly meaningful, and personal between us and among the relationships we have built—through communication, trust, and quiet understanding.

The Character of Our Heart

The natural world is the strongest and most prolific supporting source. Each of us exists *with*, not apart from, Nature. Our experiences with the lives that make up Nature influence us deeply, even at the very basic, cellular level. When we create positive experiences with life in Nature, we strengthen our sense of interconnectedness and deepen our relationship with these lives. In contrast, denial—and the many ways denial manifests—ignores, assaults, and undermines this essential need. Just as we require fresh air to breathe and clean water to drink, our relationship with Nature reflects a fundamental necessity for well-being.

Our basic need for interconnectedness is an underlying current in our lives, operating around us, within us, and through us. It connects and supports us not only with all life in Nature, but also with one another—in our friendships, families, workplaces, and communities, near and far. We cannot escape the inward and outward pull of interconnectedness.

Over time, we have evolved to become increasingly removed from the relationships that root us to Earth. In many ways, we have enriched our lives, but we have also complicated them in ways that distance us from what is most essential. What we carry in our hearts matters. Our motivations, the values we hold, our intentions, the way we choose to live, and the actions we take all shape our lives. There is no home for false intentions. Regardless of the words we express, intent cannot be faked.

Little Wren with a Big Message

We had just moved into our new home. I had been watering the newly planted Plants and Trees to help everyone become established. Even during our first year there, shortly after moving in, our yard was filled with young Birds being fed, taught, and guided by their parents. This began once they discovered the fresh water and feeder we had provided.

I have always experienced Wrens as playful and deeply curious. Three young Wrens and two adults visited our yard regularly. They moved through the shrubs and natural areas, chattering to Willow while perched along our fence. Their presence carried a wonderful sense of excitement and joy.

One day, wearing my wide-brimmed yard hat, I sat at the base of the deck stairs, taking a moment to just *be*. Although I was seated, I was not entirely still—scratching here and there and turning my head often to take everything in. I could hear a young Wren moving through the yard, exploring.

Suddenly, this little Wren emerged from beneath the deck and landed on my hat. I felt his or her tiny feet on my head. They were alone, I believe. I gasped in complete surprise and was instantly filled with excitement. The little one chattered as though I were simply another stop along the day's adventure.

Instinctively, I raised my hand, palm open, almost as if I could catch Wren if they were to fall off my hat and into my hand. They hopped onto my brim. Perhaps they were noticing my hand, which was now close by. This little Wren continued chattering for a few moments before flying off to continue their day.

What wonderful energy the experience created within me—a flow of connected energy. It was as if I were a part of this beautiful, larger situation. Perhaps this was a sign that I had been accepted as a member of their broader community.

I do not know whether the young Wren realized they had landed on me. The entire experience lasted less than fifteen seconds. I had moved and gasped when this little Bird first landed, but I remained aware of the importance of staying present and offering a genuine sense of calm. I needed to give off the right energy because life in Nature seems to know when calm is authentic and when it is forced. In that moment, Wren appeared to include me in their exploration as though I were simply part of the landscape.

Human beings often live as though they are separate from Nature. Yet there have been so many moments when, among life in our yard, I have felt as though we blended together. At times, Animals seem to either not notice me or regard me as a familiar presence, even when I move openly in the shadows. Perhaps they are fully aware of me and accept me as one life among many.

So much about this experience makes me happy.

Nature Breathes Through the Trust of Interconnectedness

Trust can make or break human relationships. It is often manipulated or taken for granted, sometimes without recognizing how deeply this harms others and ourselves. Trust strengthens bonds and builds bridges. When rooted in genuine care and respect, it blossoms and spreads.

Nature honors trust above all else. From the perspective of the natural world, human beings are not automatically trustworthy. Trust is weighed, tested, and earned.

I have been witness to the teachings of trust in the natural world. Life in Nature trusts until given a reason not to. Likewise, it seems that a lack of trust is carried with the individual who has broken it. In the same way, those who are trustworthy appear to be known for their character, as if their intentions are shared through the greater community of life.

Created relationships founded on genuine trust can have a profoundly positive effect on our health and well-being. As we become more mindful of this connection, our behaviors begin to change, old patterns soften, and our understanding of ourselves in relation to others is reshaped.

Trust can be restored, though it may take time. To earn authentic and lasting respect, we must first cultivate authentic and lasting trust within ourselves. Our intentions and what we seek from a relationship matter most.

To trust life in Nature is to recognize the inherent worth of these lives and to act in ways that support their well-being. Sometimes this means leaving them to live as they wish, and allowing them the space and resources they need to live and thrive on their own terms. It means considering their needs alongside our own and respecting their right to exist as they are.

What might become possible if we chose to restore trust?

Reciprocation

Interconnectedness might be radiant when we are children, but it is not something lost that we must search for as adults. It remains within us, waiting to be recognized and embraced. It lives in our intentions as well as in our bodies, brains, minds, and consciousness, echoing throughout our health and well-being. Interconnectedness comes alive through relationships and interactions.

Our intentions carry energy and extend outward into the world for all to "see." Life in Nature responds to this. When we allow these lives to be who they are, and when we spend time among them in a calm and respectful state—always honoring boundaries, especially the need for space—we begin to build genuine relationships. These relationships deepen when we learn to reciprocate what Nature needs.

Reciprocity matters most in the relationships we cultivate close to where we live. In our yards, neighborhoods, our most cherished homes, and communities, kindness offered and received creates a sense of belonging. We are meant to become acquainted with the lives who share these places with us, rather than thinking only of Nature as something distant and far away. These individuals are already seeking to live among us—let them be so they can be. When we allow them the freedom to do so, they are able to thrive—and so are we. When we remain mindful of this, our intentions are "heard," and this positive life-giving energy flows back to us in return.

Reciprocity is woven into all life in Nature, including our own. Our health and happiness depend upon this mutual exchange. Through reciprocity, we share our worlds by allowing one another the freedom to live according to our unified needs while coexisting together in balance.

Giving back, honoring the gifts we receive by extending them to others, is not a new concept. Yet in much of modern Western culture, we have often taken more than we have given. Historically, entire ways of life have been pushed to the edge of extinction when people were stripped of their lands, their relationships, and their connection to interconnectedness itself.

More often than not, reciprocity is of emotion, intelligence, wisdom, learning, companionship, and respectful relationships grounded in healthy boundaries.

Perhaps we can soften our voices, decide to listen more deeply, and allow Nature to experience reciprocity within our presence. We can step back quietly and learn from these lives rather than deciding for them what they need.

Expressing Gratitude

It was a very cold day, and I saw young Squirrel trying to get to the feeders. So I went outside and placed two walnuts on the post, then put out more seeds for Birds.

I had usually seen this young Squirrel run along the wall back to Trees near our home, but I did not see that happen that afternoon. After putting out the food, I went back inside.

About an hour later, I looked out the window and noticed a gift resting on the same post where I had left the walnuts. These two young Squirrels, who normally moved stones around throughout the day, were nowhere to be seen, although a stone was left. They were likely staying low and conserving warmth. I felt certain that this beautiful stone had been left as an expression of gratitude.

I had experienced this before on cold and rainy days—larger, heavier stones than one might expect such a small being to carry. This young Squirrel continued to express gratitude in this way. I have a small collection of these gifts now, each one a tangible reminder of moments and relationships that will remain very special to me for the rest of my days.

I have yet to witness the gift-giving in the act, although I once saw one Squirrel, who was especially dear to me, holding a stone in her mouth as she watched from the porch. Movers were carrying boxes from our home, and we were leaving the next morning.

The sentiment behind these offerings—the reciprocity of gratitude—has never fallen on deaf ears with me. There is such warmth and expressive language behind this quiet act. This was real emotion shared through interconnectedness among lives on this beautiful Earth, our home. I have rarely felt appreciation expressed as profoundly with acts of kindness in moments like these, where no words were spoken.

I miss everyone so much, even years later. They transformed a house and the land around it into a true home. They were and are my home.

Eleven

The Time of Them

A Sacred Understanding

There was a time when lives in Nature were known as *them*. These lives were known with inherent worth and were appreciated for their unique gifts and intelligence. They were valued in their own right, and learning from them was revered and essential. They held their place in the bigger picture, alongside humans. There was balance.

Then, as change occurred, with many acting in ways that spread disregard, these lives—and Nature—became an "it." When he, she, you, and them became "it," the purpose of their lives changed. As a result, *life*—including people, our mental and physical health, and our attempts at *balance*—has continued to decay and suffer.

Our relationships with others are greatly impacted by positive and negative wording. The treatment of others is contingent upon how we refer to them. Through interconnectedness, there is a change in us, in those we are considering, and in anyone else in our lives. How we define others depends on us. We hold the solution for positive change—one that ripples to all life around us and within us—returning to our larger community of life on Earth from a state of "it."

There was a brotherhood and sisterhood—a respect for the depth and personality within life throughout Nature—and for what these lives, including their intelligence, offered to people. They were sustaining, and this did not fall on deaf ears. These lives were valued and appreciated, and there was an understanding of how important gratitude and reciprocation were in sustaining all life. Reciprocation is not among humans alone.

The original people of these lands live closer to Earth, weaving through all life with a sense of sacred understanding—living as one. I do not have the right to speak for them. I do know that many still live in ways that support health and balance with all life in Nature, and have done so for thousands of years. Then, an alternative way of living spread like rot across these lands. It did not have to be this way. But it continues, still.

Harsh words fit the harsh way we treat life in the natural world. Hold honor within. Life in Nature is all life on this planet. We are not exceptional without the trust and respect of all others—earned and supported through complete and mindfully balanced relationships.

Touch Earth

The body does not forget Earth completely. As far as time is concerned, this understanding was not that long ago. Perhaps, if we appreciate the dignity of lives that honor all life in a way that focuses on balance, then in healing moments, we too can embrace the value that has been missing for a very long time.

We cannot be complete without being connected. We do not need to forgo all we have accomplished. We can do better, create better—be better.

Many are searching for more or feeling lost with less. Enter our brilliant natural world, and our enduring need for balance among them.

Honoring what was, and seeking ways to embrace what we could be together—with trust and respect—is the start; a simple and profoundly beautiful act that would not take much of us on a daily or even weekly basis. In just a moment, we could shift the energy left by those without compassion or care, and begin to feed acceptance, kindness, and unity into the equation—with and among life in the natural world.

Unless native to these lands, all of us have origins and ancestors who traveled from other shores. In a very short time, so much harm has been inflicted and so much good has been torn away. Taking too much, taking advantage of good, and taking for granted any goodness offered will overpower any good way. The negative creates more unrest within our natural self, and we take more in an attempt to rid ourselves of that discomfort.

But change must benefit the whole. We, too, are *of Earth*.

Fluid Circle

Our world is very different now. These relationships have healing to do, as we have broken the trust. We must own our part in these broken relationships and do our part to meet life in Nature halfway. We have done so much to scrape away all that we can, then move on to the next, without considering the harm we have done—let alone what it might mean long term.

It is time to come back around—to circle back to what worked so well—and begin to embrace life in Nature again. We can take all the good we have come to know and have created for one another, and include all the good that exists and was created for all life in Nature. For some of us, this means beginning to build our relationships with life in Nature—with one another, and with the life around our neighborhoods and homes. We need it. Nature needs it. Our Earth is telling us we have not been listening—and now we must change, or change will happen for us to restore balance.

Change starts and ends with each one of us. We are very capable within ourselves and within our own worlds. We can find great wisdom and guidance from others—listening with open ears and a closed mouth. We can evolve without dishonoring any of our personal beliefs, when rooted in kindness.

Interconnectedness runs freely and is not bound to human terms of belief. Kindness is kindness. Hate is hate. Violence is violence. Nature needs us to stop destroying. We need Nature so that we are not broken, suffering, ill, or destroyed. If our hearts are good, life in Nature will know. If our hearts are not good, Nature will know that, too—and we will carry it in countless ways.

We need to return to a positive evolution within humanity, and it must include honoring our interconnectedness with them—the life outside your window. Our immediate change will reverberate throughout all life in Nature. Some of us have not forgotten this way of being. This is our empowerment.

It is interconnectedness that holds space for all that we are—and who we are, too. We are never alone—look to all the individuals found in Nature. We are part of the same community.

Love Trusts in Silence

Little Squirrel came down Tree, to the feeder. I startled him or her, and the little one ran up a bit. The feeder was full and clean, so I stayed still, and Squirrel came back down and just looked at me. Wanting to offer respect and hoping to convey a trusting environment, I moved softly from the corner to the other side of Tree, about three feet away. Squirrel left the food and came around to my side, holding onto Tree, and stared at me.

I calmed my heart, felt love, and walked closer. I do not usually do this, as space and distance are of very high regard. I felt pulled to do so in this moment—independent of me, not of my wants or needs. We were about a foot from one another. There was a need to look into each other's eyes.

I never like to stare or look directly into eyes in Nature because of the fear this language can create. It is interesting, since this is encouraged among people—along with a swift, fast, and often obtrusive walk, a "confident" walk—something I never do when I am in my yard with life in Nature. I walk softly, calmly, slowly, without stalking—respectfully. The walk of a "confident" human, as defined in human terms, does not go over well with life in Nature. In Nature, there is no need to declare who we are in these nonverbal ways. Who we are is already known, because we radiate it through our senses. I believe it is carried in our hearts and in our intent.

Lately, I have noticed that Squirrels are less fearful and seem to be okay if I look into their eyes, even if only for a brief moment. They stare and process information in a way that fascinates me—something that goes beyond my love for them, reaching into the closeness I longed to feel and understand as a child. Squirrels were the first individuals I tried to connect and communicate with as a child—not with words, but in ways I felt, though I could not explain in thought. I love these moments, because I understand now that it has always been all that I am *within* myself, reaching for interconnectedness with the natural world.

In that moment, we looked into each other's eyes for about thirty seconds. We were quiet, with no spoken language from me and no sounds of Nature from Squirrel. I felt love. Even in reflection, it brings tears to my eyes, as I can return to that moment so quickly of deep, true love, respect, and trust. With this comes peace, calm, and balance—all through the senses. What was created together in that moment makes me feel empowered even to this present day. The one thing I have always wanted in my life is to communicate with Nature on their terms, in their way, in their world. Beautiful.

I thanked Squirrel quietly, from within me, without spoken words, and slowly left them to eat.

Twelve

From Them to "It"

"It"

Since words build and words destroy, and human vocal cords—language, words—do not exist in Nature, we are left in our own human world of verbal communication. When done well, it is done well. Life in Nature, with all their varied forms of communication, understanding one another is with fluent ease. But we are left alone on our own island. We have dominated a few senses, along with accompanying emotions. Removed from the equation, we are removed from the conversation.

We must understand *now* that how one refers to another creates the ground upon which the other is treated. This is clear proof of the power—the energy of intent and motive—behind words. When someone is not viewed as an individual—a sentient being—with intelligence, emotion, experiences, and contributions of their own, but instead as a thing, an *"it,"* they become an object no longer seen as worthy of fair and justifiable treatment.

In these words, we become even more divided. How we treat them says a lot about where they stand within our ethical framework and worldview. An "it" is automatically made inferior and therefore less important, regardless of how they perceive themselves. As such, we treat life in Nature as large groups of "its," like old tires sometimes left to waste or gathered in large numbers—often in uncomfortable, painful, and intolerable conditions—for our own personal use. We do not want *it*, yet we do not care to change ourselves to change it. Because IT starts with US. We feel nothing when we leave the tire behind, without regard. It is nothing to us.

The words we choose to describe life in Nature say more about us and our attitudes than we might think. Our chosen words create the stage for how we view another, and this greatly influences how we treat them—and what happens within us. What we think, do, and are, as people, directly impacts our mental, physical, and emotional states—always. Our health and our happiness, the words we use to explain who we are, our situations, and our interconnected state, have a profound impact on our emotional and physical well-being—our inner landscape.

All Birds are not the same. When we learn from them—when we take the time to acknowledge the beauty and complexity of the individual—we begin to understand uniqueness. When we welcome moments with individuals, on their terms—those who might frequent our yard or cross our path on daily walks—we begin to truly know them. The area around our home, our neighborhoods, and our communities becomes more familiar...a place filled with life worth protecting, for the good of the whole.

To assume life in Nature—from Cows to Chickens, and from Yellow-shafted Flickers to Great Horned Owls—are merely *things* invites us to reflect on how we treat other people as well. Our words shape our views, and our views shape our beliefs. We are all—each one of us—unique individuals, living together within groups, striving for healthy lives on this shared planet Earth. There are no "its" in Nature. Life in Nature does not operate in human words. We might learn a thing or two—or fifty—about trust and respect from the lives around us each day. And it all begins with how we refer to them.

As words are powerful, and too often misused, we can start harnessing that power through our own intentional use of words—in how we refer to ourselves, our family members, our coworkers, and the lives quietly waiting to be known by us as well. When we do this, we draw to us all that we are and all that we send out into the world. To build a bridge of trust between them and us, begin to call these beautifully complex and emotional lives by words that honor all that they hold. Begin to train yourself to release the use of lifeless "it" and "things." We have had it wrong—but we can make it right again. And that power rests within each one of us.

Our words can build us up and empower those who take an interest in us. Rooted in our motive and intent, our words create an energy that lingers in the air around us. They can also tear down—inflicting indifference or pain—revealing who we truly are to the world around us, to one another, and to all of life in Nature.

In Our Powerlessness, We See Defenselessness

It is our own perceived, self-punishing inadequacies that do us so much harm. Regardless of how these images of self come into our awareness, we are the keepers of our own domain.

There are some things that are ours, of our own making: our education, our experiences, and our personal integrity. There is little that another can touch in these—what we have learned, what life has taught us directly, and what we hold dear within our hearts. When we grow to feel powerless in our situations, we can call on our education, all we have learned, and our integrity to help keep us centered. For those who remain in cycles of pain and hurt, they may lash out or manipulate because, in some way, it brings temporary relief within their misery—because they are unhappy, and therefore unhealthy, within their physical bodies.

When we choose to move through life in this way—from a place of inner defeat and powerlessness—we begin to see and approach others from a position of defenselessness. And although harm may come easily through physical strength, proximity, or size, the energy set out into the world is one of insecurity and weakness. Our health is impacted.

We become powerless to our enabled insecurities, which then impact all of our relationships—including the most important one we have with ourselves, and how we view our own health. Because life in Nature responds to positivity, just as it does to negativity, these relationships feed our interconnectedness, and the damage can be profound. What we do to life in Nature, we also do to ourselves.

So when we refer to anyone as "it," we effectively remove the potential for emotional connection or recognition of the worth that belongs to that individual. The "it" becomes an object—subject to our needs and assumed to be at our disposal.

This is limiting, to say the very least. Our lack of consideration for others who share this world limits our openness and our ability to accept. And so, we become victims of our own insecurities and powerlessness.

When we choose instead to value another for who they are—and for what they hold within as their own worth, even if we do not fully understand them—we open doors for growth within ourselves and across our relationships. This is the foundation of our personal power; a support for the positive intent of our beliefs—our unique greatness, just as it exists throughout life in the natural world.

Life throughout the natural world is full of thought, feeling, presence, values, drive, and even playful love. They do play, and they do love. We may know how to build microscopes and study cells, yet we would be lost in understanding the needs of soil and the lives within it that sustain and support us. We cannot build a web of both sticky and non-sticky strands and remember its design well enough to move within it ourselves. We would struggle if left to travel great distances outdoors, navigating weather, danger, and the search for safe places to rest or live. We cannot see ultraviolet fields or other forms of light that guide so many. There are lives in Nature who work tirelessly to protect those they raise, who feel deep loss, and who persist in keeping their young safe from harm. Regardless of how different this—and they—may appear to us, without taking the time to truly understand, they build, adapt, defend, and act for the good of the group, often discouraging those who act only for themselves.

Regardless of size, mobility, or lifespan, if we see so little inherent value in lives in Nature, does that not influence how we view other people as well?

We have become so removed from our inner world; our *within place*—from our thoughts, minds, and bodies—that deeper care and concern for these individuals in Nature is often seen as weak or childlike. Without this trust in ourselves, this natural care is dismissed, especially by those who are even more disconnected. These relationships are not silly, and they are far from unimportant. In truth, our relationships with Nature may be among the most important we have—foundational to all others.

Look at the state of life—of people, of health, both mental and physical—and where we stand with balance. The words we choose to use when referring to another create the environment from which the relationship grows. This holds true for our relationships with people and with individuals throughout Nature. Words leave us as energy, and that energy has an impact on us and on others. The way we treat others is shaped by how we refer to them. How we define others depends on us.

We hold the solution for positive change—one that ripples through all life around us. By changing our words from negative—even those within our thoughts, shaped by the deeper currents of insecurity and fear—something within us begins to shift.

Why See an "It" When Not a Thing?

Referring to someone as an "it" places them at a distance from us, allowing us to align with our current values and morals and treat "it" as we wish—without impacting how we see ourselves or what we believe we need. We rationalize our treatment when someone becomes an "it." "It" is assumed to act only in response to stimuli, living by instinct alone. An unaware "thing," reduced to instinct, is seen as indifferent to what is happening around it—or to what is happening to it, its family, its young, its community, its environment, its migration routes, its waters and oceans, and its streams. No feeling. No fear. No pain, suffering, or grief.

If Robins, Songbirds, or another individual Bird flies into glass, it is dismissed as "dumb"—not recognized as a consequence of glass not existing in the natural world. This is not understood as exhaustion or hunger after traveling unfathomable distances, waking in an unfamiliar place. Older Birds are aware of glass. I have watched them—and I have watched younger ones learn the hard way.

Plastic floating in the ocean, mixed with food, is not consumed out of stupidity. Plastic does not belong there, and a floating bag can resemble a food source—something that should not cause harm or kill when consumed. With all of the archaeological wonders left behind on Earth, we are leaving behind toxic structures and vast landfills full of crap that none of us can avoid accumulating.

Waste that will remain. A polluted inheritance—our imprint on history. And it often seems to be left without regard or consideration.

Fortunately, within this reality, our greatest weakness also holds the potential for strength—if we choose to see it. A larger, more complete community stands ready to teach us, if we are willing to learn. They wait. Because we are not doing so well. And still, as a species, we refuse to see. Why?

Now is the time to start by changing the way we think. It may take practice. We may slip, again and again, referring to Tree in our yard as an "it," or Dog as an "it." But we can retrain our minds. Unless we choose not to try, this is a change that improves us—and all life around us. Its impact is far-reaching.

Our attitudes impact our health, whether we acknowledge it or not, no matter how much we try to run from this fact. We need Nature. But first, we must create the conditions—the environment within ourselves—for this relationship to be recognized and restored by those with whom we seek reconnection.

Thirteen

Trusted Because We Are Trustworthy

The Call of Crow

Crow was across the street and appeared to be looking into the window where I was doing yoga. I opened the blinds and window to see him or her more clearly. He or she was still looking at me in the same way, calling now and then, calling in another direction, perhaps to another Crow family. Then, Crow jumped over to another roof right next to where they had been, still calling.

Then, on Sunday of the same week, my husband and I took a walk. On our way back, and not far from our door, Crow was carrying nesting material and flew overhead, off to the side where I was walking. I saw some Moss in their mouth as they stopped flapping their wings and soared, looking down at me. My husband was in front of me and over a bit. Crow was right above my head and made one small soaring circle above me before continuing straight on to where they were going.

A quick moment later, a second Crow arrived silently, landing on a roof within view of me, watching. Quietly looking.

I could sense something in the stillness.

I walked to our door and went inside, grabbed a handful of raw, shelled, unsalted walnut halves, and went back outside. Crow was on the same roof, still within sight of me. Calmingly, I went to where they were and placed the walnuts at the base of small Tree. I turned and headed back toward our door. Crow dropped down, gathered the walnuts, and flew off with them.

This was my first time meeting these two individuals. I felt recognized but did not know how, since I was still so new to the area. Only a few individual Juncos and Hummingbirds knew me, or so I thought.

About a month later, I saw two Crows on a roof on the other side of our house, just as before, but they had both landed at this other spot. I watched one Junco fly up to one Crow, and Crow looked at the individual in a very brief exchange. Junco flew off, and Crow left to come around to the other side of our townhome row. He or she soared above and dropped lower to the spot where I had left walnuts only that one time before. They were not interested in more. When I went back out, they watched me approach. I stood back, and we looked at one another. Then this individual left. They wanted to see me and nothing else.

On several occasions, I was in the right place at the right time to witness Juncos and Crows work together. I was certain they were helping one another here, too. They guided one another to food, water, and locations of safety. I believe the same is true with Squirrels—they create an order and timing to visit our small yards. Crows and Squirrels work together to protect and watch over young and help with food. They, too, understand and focus on individuals they have come to know.

Something else is going on that maybe we are not meant to fully comprehend. I trust in this, though, because that seems to be the point—trust.

This exchange with Crow reinforced to me that my level of integrity with them was the focus. Trust is not built on words, but on recognition and respect.

We Want to Trust

In trusting relationships, life in Nature holds very strong bonds. Trust is woven into their way of life, how they meet and greet one another, and within their close relationships—they have never lost touch with it. In these trusting moments, communication is conveyed: each life might share much more than greetings, gratitude, affection, friendship, and love.

Where we have broken trust out in the world, we can acknowledge our responsibility, apologize sincerely, and make things right going forward. We can do things differently. It is through the support of being with Nature near our dear homes that we will find our own way to trusting who we are and, therefore, the right others have to rely on our trust, too.

It is interesting to me that trust appears to be rarely discussed with children. Instead, we hear about trust in discussions or through moments in phrases such as, "I don't trust him/her." Trust is not really explained to us. It is as if knowing and understanding trust is a given. But is it?

The knowledge of trust is more within. It is felt and known—sensed. People can try to manipulate it with words, but eventually, they fail.

Trust comes from the same source as joy and love. The function of trust may originate from our parents, our siblings, our grandparents, or our friends, but this is not always true. Where do we learn trust? Where can we breathe it in? For many of us, trust has a very unstable start, and this instability might still be following us. It can have a great impact on our lives. Yet, we have a natural sense of when trust is present and when it is not present in our interactions with other people. But we can grow to ignore these strong sense reactions, especially when trust is manipulated intentionally in some way.

We *want* to trust. In moments, we ignore that feeling when something feels "off." We can ignore the high value and relational need for trust. We can ignore self-trust. We can turn our backs on the trust within ourselves, souring inner trust.

Our motivation and actions impact us and do make or break trust. Our inner Nature includes our physical, mental, and emotional health, and these parts of self are interconnected. Our inner Nature is intertwined with our attitudes, motivation, and actions toward life throughout the natural world because of our interconnectedness. As members of this beautiful planet Earth, our reason or motivation for doing what we do in life has an impact on our health and happiness.

In Nature, humans seem quite comfortable taking advantage of trust. Life in Nature knows better the role of motivation and intention and how vital trust is to relationships, happiness, and survival. There are numerous ways to know trust, but we seem to have forgotten the *knowing*, or we ignore it altogether.

A lot of natural life relies heavily on the sense of smell, a sense in which, even at our best, we are often far inferior. Those with highly attuned senses of smell can better understand others as well as their immediate environment and what another's presence means. Scent sends messages to many in the natural world. Perhaps trust has a scent. In Nature, trust is part of the process of understanding the intention of the relationship. With Nature, our motivations are quickly known.

Trust Is Sensed and Known

Trust is sensed. Trust is unparalleled in its importance in relationships between people, within our sense of self, and with life throughout Nature. Our basic need for interconnectedness heavily holds the sense of trust.

Trust itself is intangible. We cannot offer up a block of trust to anyone. Trust is the air that flows through all our interactions. Like blood is to the body, trust is to a relationship. When in a trusting relationship, the tangible is created through our senses, emotions, thoughts, and all the physiology that follows. Trust is an acknowledgment of the value we hold within ourselves and with the one we greet in our interactions.

It is easier to understand trust's role in human relationships. Many of us have had trust broken, manipulated, abused, or ignored, so we understand what it feels like to trust and not to trust. The same holds true with life in Nature. These individuals place trust front and center. For people, we often wait until damage has been done.

The need for trust is within all. This makes perfect sense when we remain aware that we are all of this Earth, together. We are more alike than not. All of us know that we want trusting relationships.

A whole world of communities across oceans and seas, deserts and mountains, meadows, and forests exists where individuals function without human-induced egos.

When our motivation is to care, we are caring individuals, truly and without manipulation. Trust finds deep roots. And life in Nature knows and senses this, too.

To mend broken trust and to build healthy trust within our lives and with life in Nature, we need to care truly for these lives, starting within ourselves. We must care about who we are as individuals and what we radiate into the world. To care in healthy ways for others, we need the same source from within ourselves. We need to care about who we are in healthy ways, too. We need to start working on caring, and if we struggle with caring for those beyond people, we need to look within our hearts as to why.

It is time to expand and broaden our thinking to include the individual lives that call these areas home, too. We need to take into account their needs and interests in our neighborhoods and communities, for it is here where we will be most supported, balanced, healthy, and inspired. We are seen and known here with the lives that join us.

In Nature, we find ourselves in our truest light. We begin healing. We have to heal these lives, too. One cannot happen appropriately, interconnectedly, or completely without the other. We need to trust in trust.

We know that our relationships with life in Nature, and how we approach them, have a profound impact on our physical, mental, emotional, and personal beliefs. Our bodies are designed this way, just as they are throughout life in Nature. We must alter our thinking of what we want and need to also include what life in Nature needs.

Interconnectedness Through the Vein of Trust

Trust is the invisible foundation of every relationship—the soil from which connection grows and interconnectedness flows.

With life throughout the natural world, trust is known and matters. Here, the breeze blows softly through the leaves. Little Insects dance in communication with one another through rays of sunlight, quietly and peacefully.

There is something to this—these interactions. There is more to our relationships. Interconnectedness is always present within us. The need for interconnection has always been within us, and trust brings interconnectedness fully alive.

Trust is the link that connects our natural self—our physical, mental, emotional, mindful self—with all life beyond our physical state. Our senses, and where they overlap, are our common language. We need interconnectedness. We need to be open. We need trust.

Our actions and kindness reverberate within us and extend to other relationships. Our kind acts do not go unnoticed. In my experiences with individuals in Nature, my actions have followed me across the country, over mountains, and through many homes.

We breathe with interconnectedness. Yet, interconnectedness will not extend to us in our attempts to capture images we believe other people need to see. These moments are of such deep worth; they are meant as gifts for us. We can decide from this day forward to honor, accept, and learn from the offering. Trust does just that within the interconnectedness of experience.

We can either honor trust or use it. We are known in Nature quickly. In these trusting relationships with the natural world, trust is reciprocated. Sometimes it takes time to build, and sometimes it is broken. Have you broken trust with someone for whom you should have always offered a safe place to be? If you are fortunate enough to have deeply trusting relationships, imagine your life without them.

Trust in Bear

We tend to have our hands full trying to understand ourselves and our families, not to mention humanity. It would be inconceivable to know everyone or even fully understand the complexities that surround particular species.

Bears, for example, are often misunderstood, like so many individuals throughout Nature. What is important to remember is that Bear has a life, a world, needs, family, and goals for each day. They are complex individuals with complex lives, engaging in complex communication within a complex community. Allow them to be who they are in their world so that we might be okay in our worlds, too. They are someone. Trust in this. Trust in them to live well.

Standing in the Light of Trust

Trust is in the air, moving around us, flowing between and through us from our *inner Nature*. Trust is not forcing contact on another who wishes to be left alone. It is not capturing and containing an individual life who is not meant to live confined and alone. It is nowhere to be found when *invading* another's boundaries or space. Trust is present when we uphold space for boundaries. In all of our relationships, *we become* the trust that is present between us.

Some of us might be very new to standing in the light of trust. Still, we understand what trust is and means, even if our lives and choices indicate otherwise. I believe we are never alone, even if all of our relationships with other people stomp on our individual needs around trust or exist without much trust at all. By going to Nature in trusting relationships, we can know and feel the support we need. They have not lost this way as we have. They are masters at offering healing through present states of calm and trustworthy moments.

Core confidence is restored when relationships with Nature are developed over time. Therefore, our trusting relationship with ourselves is also restored and developed over time. We feel balance. We feel peaceful, calm, and relaxed. Our headaches and heartaches diminish. Over time, when we not only understand but *know* that we are part of Nature, our health can improve considerably, mentally and physically. Positive creates more positive. When our confidence in self-trust finds common ground in kind moments with individual lives in the natural world, we also begin to value trust in other moments with other people. We start to place high value on boundaries, trust, and seek what is best for the individual without harm.

Our gift of progress is missing our relationships with Nature. And not Nature as a whole, because it starts with individual life close to home—in our yards and communities, among individual people learning from individuals in Nature. It begins with learning objectively who they are, what they need, what they wish to share, and honoring their boundaries and space to be, just like we need.

Trusting relationships develop. Kind regard and consideration to leave them alone offers great respect. In person, and with proper distance and space, real information is passed and received. Relating, connecting, and the experience of the senses—this is interconnectedness.

It starts with us. Our greatest strength is sourced from within. It is our inner support for who we are at our core, without facades or false fronts. Doubts and fears fall aside, and we can rest more easily in moments of uncertainty.

Trust in ourselves—this source of inner support—allows us to stand on our own during moments of doubt and know that, in the face of anything, our trust within can never be taken away.

Building our sense of self stemming from self-trust is our strength; it is a part of who we are naturally as individuals. The stronger we are in this area, the easier we can be in the presence of life in Nature, within our truth. This is interconnectedness fed by the nourishment found in moments when we can be in the presence of others—nonhuman and human—and offer space for these lives to simply be.

Squirrel with Stripe on His Nose

We moved in during the month of June. This was when I learned that Squirrels vary in their relationships with me. The younger ones—or those who appear younger, based on size—keep a distance, but watch me. They watch closely, still as can be. I look away to offer space.

There are older individuals—Big Mama Squirrel and a male with a stripe of darker fur running across the bridge of his nose—who seem aware of my intent. The individual with the darker stripe of fur became known as Stripe.

I met him when he would hide in small Tree and eat from one of the cups for Birds. At first, he would back deeper into Tree, but then he began staying where he was eating and simply watched me if I walked up to check seed levels.

He learned to stand on one of the fence posts by the birdbath and wait for me to see him. He waited for food by the post near the steps leading to our deck. He came onto the deck and stood on the railing to look into our windows. He did not always seek seeds from the feeders. Sometimes, he seemed to look in—to understand.

Other Squirrels began doing this now, too. This was new to me. A few Birds started doing so as well. There was one Grackle who came to the window to look into my world.

When I headed out onto the deck, Stripe excitedly ran down and over to the post at the base of the steps. When offering a few unsalted, shelled walnuts, I tried to show him how two halves fit together so he could carry them more easily in his mouth. He watched from a distance, but trust had deepened. Then I placed the nuts together as they would fit into a shell on the post. Picking them up was different for him. He could not see straight in front of his face because of how he was designed, so he grabbed one half and then another and struggled to fit both into his mouth. This day was no different. He fumbled and dropped one on two different occasions.

The last two walnuts I brought out for him felt different. I was trying to show him how they fit, and he seemed more in a rush. He climbed onto the post and took them from my hand as I held them together neatly, as they fit. Once his mouth was around them, he reached out with his hands and gently guided them farther into his mouth. I knew it was because this was how he needed to learn what I was showing him. He stopped, nuts in mouth, looked at me, and then went off for the evening.

Squirrels are brilliant, like all the rest, and they find ways to eat from feeders. I left them a few shelled and unsalted walnuts in moments. With Stripe, there was something different about our relationship. We learned from one another in ways I have trouble putting into words. He learned with me, about me, and through me firsthand. This was our relationship.

I will never forget what I felt around him—what he taught me and shared with me through our experiences. These moments are deeply personal and meant to be held within, respectfully.

I will share that we also learned about distance, space, food, and when there was no food because it was time to put it away. There were days when food was unavailable. They were respectful of that, too. Squirrels can be very determined. But they give me space to exist as I am. Ever since Lightning visited, balance was answered when it was in need. This remained true when we moved across states and lived in different homes. Trust rested easily in balance. Life in Nature worked it out and included me and my yard, so it seemed. We trusted in one another. This knowledge seemed to be shared. I made no attempt to test it and offered only reciprocation.

Fourteen
I, Alone

Now Is the Time

Perhaps we have lost our worth in the equation. It has been lost, but not forgotten.

There cannot be a time in our lives more important than now. Now—our present. Now, we can decide each step, each day forward, each action, and each thought, from this point forward, as we face the impacts of our ill-planned and impulsive reactions to our relationships with, around, and through the natural world.

Balance—*change*—begins within each one of us. Who we are and how we feel is our choice. Why have we not noticed? Do we feel that we do not matter?

Because life in Nature lives in a vast network of senses, intelligence, emotion, and feelings—because they cannot say, "That hurts, please stop"—does not mean they do not feel pain. A lack of tears or human cries does not free them from feeling. Expressed in energy waves, chemical signals, nonverbal communication, a kind of knowing, their own voices—screams are emitted, and fear is palpable. Life in Nature feels.

Plants react to being cut down or torn away. Lives struggle to free themselves, working to break free from impending doom; the calls and yells, uniquely their own, are there. We need these lives to sustain our own. It is time to reconsider how we go about this so that these lives are offered the respect they deserve, too.

If we believe that Nature is unworthy or that our relationships with them are impossible, perhaps we need more time in front of that mirror.

It is becoming more known that numerous Animals have unique names assigned by their parents, and that they respond when called by individuals they know. This life around us has awareness, love, concern, and care for one another. They feel, and they suffer from pain and from loss. Our potential is harbored in the trust we are afforded by Nature, not the other way around. This is where strength lives. And there is no room for power-hungry individual interest here.

Lives throughout the natural world understand all this. We need to learn it again.

What We Do to Earth, We Do to Ourselves

Nature can be like the abused child: no voice, no support, deeply damaged, and left with lasting effects. We need to treat Nature differently.

Life in Nature is not to be acknowledged only when serving us or when we need something from them. Nature is not built for us to use. One's attitude about Nature reveals a lot. Sometimes we need to step back and allow another to be and to be heard.

Life in Nature is aware of inequality and, when it is inflicted upon them by humans, they too will make it known in their own way and in their own time. What we hold within and who we are with others creates the physiological changes and biological reactions that either help or harm. Physical pain, mental suffering, emotional distress, and grief inflicted upon life around us—coursing through soils, waters, air, and land as toxins—end up within us, too. We consume this suffering, and our cells take it all in.

This energy of interconnectedness feeds our brains, our cells, our bodies, our minds, our spirit—who we are. Once the roots deepen and are nourished, this ripples into other areas of our lives. We seek similar positive, supportive relationships from others. Our basic need for interconnectedness becomes second nature to us—to our Nature. Frustration and struggles slip away, and we evolve, together with Nature on our Earth.

When we are with Nature, Nature can sense our intent—our true self. Nature will mirror this back to us through reaction. When we are with Nature, we are with ourselves in the most honest way. Through this mirror, we see our true self—in pain or confusion—or in the formation of purpose and dreams. Within Nature, we can feel the most honest version of ourselves. It starts with us, alone. What are we missing? Is it possible that the one thing we are missing is us?

I, alone, create my happiness. I, alone, create harm around and within. I am not merely reactive and, although my body will react, I can be proactive. That is a choice that I, alone, can make.

Male House Finch

It was winter, and I was about to leave one state to move to another. It was cold, and there was snow on the ground. I noticed a male House Finch sitting below the feeder in the snow. I went outside and tried to place him up in the feeder so he would be out of the cold snow. He flew back down, so I sat with him outside. I held him gently in my hands so he would stay warm.

His breathing was labored. He was closing his eyes slowly. I knew he was dying, maybe from age or from the cold—I do not know. I did not want him to die alone in the freezing snow, so I stayed with him. I did this very quietly and calmly. I did not want to stress him even a little. If he had wanted to be alone, I would have known, and I would have left him in peace. I stayed calm and peaceful. I was outside with him for at least forty-five minutes. His condition remained the same as I stayed with him gently, giving him the space to return to the snow if he wanted. He chose the warmth and rested.

All of a sudden, his breathing changed, or possibly stopped, and he seemed alert and sat up. He seemed like he was going to be okay, looking around as if he was taking in the surroundings and getting ready to fly off into Tree right next to us. He sang his song one last time—a beautiful, loud, full, clear song and voice. His wings stretched out as if preparing to take flight. His song ended, his eyes closed, and he pulled his wings comfortably back to his sides and rested in my hand. He had moved on, transitioning from our beautiful Earth to what seemed like another reality or place he was happy and ready to join. I felt so much from this experience. I cried a lot. It was very emotional, beautiful, and freeing all at once.

I am convinced Birds, and all life in Nature, like humans, see another dimension, world, place, or life and enter it once they leave this Earth. He was happy and sang as he raised his wings and his energy moved on. He saw a place, something, and was very happy to go to that reality. His physical self reacted as his energy transitioned and changed.

It is a very special place to be when sitting quietly with an Animal, an individual, certainly any loved one, as life changes energy form. I believe the energy moves on to another place or dimension and can return and remain alive around us in some parallel place—comfort when needed. There are words for this place. Perhaps what matters most is not the labels assigned, but the profound love itself. In traumatic situations, regardless of who they are, surrounding them within our hearts with kindness, unconditional love, and support matters deeply. Our sense of love always keeps us interconnected. In our quiet moments, and this includes meditation, we can enter a calm, silent space that only gentleness and harmony can create, and we still find a way to be together.

To be with someone at this point in their life is such an honor. This moment with this sweet little Bird remains with me always. I have only experienced such a moment one other time.

The Oneness of Me

Since we are interconnected, I alone can make or break these relationships, single-handedly. I can address my struggles through the support found with life in Nature and through the alternatives available to me all the time. I can do my part to mend these relationships and create a new path forward. I alone must build, rebuild, or evolve my relationship with these lives so that I can establish trust—to prove that I am worthy of their trust. Allowing another to be who they are, honoring their rightful boundaries, space, families, and needs—these are worthy of trust.

We support one another in this relationship and in our struggles, but we must be present. Being with Nature as a trustworthy representative of humanity offers stable ground. Nature waits silently.

Our senses and what we feel through them are like the light and fresh breeze in the house of who we are in this world. Our yards and neighborhoods can provide unconditional support and care. It is up to us to take ownership, accept our feelings, and honor our emotions so that we allow room for the acceptance of others. History has shaped each one of us. Single-handedly, we create our happiness.

So many signs and signals for connection from individuals in Nature go ignored. We need to make the first step. We need to stop and listen. No one can make us know or believe in an experience unless we experience it ourselves. Our attitudes, choice of words, thoughts, intent, ability to remain open, willingness to set aside assumptions of what we know, what we have learned, what we have believed—so that we may see what Nature has to share with us. Without pressure or dominance, but with gentle support, we are offered a glimpse into a world we assume we know when we really know so little. Again, it starts with us. Nature watches quietly.

Little Female House Finch

She flew into the sunroom window, so I went out to make certain she would not suffocate with fragile lungs while lying on her back. She was on her back trying to fly. I startled her…terrified her. She yelled and yelled as I tried to pick her up while she made every effort to fly from the ground. If I did not get her upright and off the ground, she would freeze and suffocate under her own weight. Birds cannot survive being knocked out—which happens to so many—or lying on the ground unable to prop themselves up, such as this little one. The window strike stunned her, and she was unable to fly at that moment, though I hoped she had no injuries that would prevent her safe return to Trees and back to her family.

I was able to get her into my hands. Her heart was racing, and her head did that strange ticking movement as I rotated her gently, as though she were trying to focus on one thing and then the next. She was trying to balance. She had hit her head directly into the window during her flight. She appeared dizzy, perhaps experiencing something similar to vertigo. I stayed near Plants because I sensed that they made her feel safe.

Birds like the cover of Shrubs, and I knew from the recent late-May heavy snow and freezing temperatures—and from the number of Birds pressing against our home for warmth—that she would be familiar with that area of the yard. After the snowstorm in May, many stayed close to those Shrubs, sitting on our outside windowsill with their chests pressed against the house. Many were soaked from snow clumps falling from branches during the morning melt. I could only assume they were desperately trying to draw heat from the brick into their tiny bodies. It was unsettling and difficult to witness.

So I sat beneath those Shrubs. I tried to see if she could fly because then I could place her safely on a sturdy branch until she felt better, but she could not. She dropped onto her back as I released her close to the ground and tried flying that way again, screaming—confused and terrified. I feared she had broken a wing or injured her neck. She screamed and screamed, panicked and so frightened. Other Birds came over. A male Robin came to see what was happening and almost landed on my head. Male and female Finches gathered above us in nearby Tree on the lowest branch, chirping to her as she cried out in fear. I stayed very calm and relaxed throughout my body so she could feel that calmness too. I was able to settle her comfortably in my hand, and she stopped yelling. Her heart was still racing. I kept her head upright and let her rest. Finches stayed close and called softly to her. She calmed.

I sat there for at least fifteen to twenty minutes. Birds around us gradually returned to the yard to feed, drink, rest, and behave normally. I did not remain perfectly still. In an effort not to startle them or make them think I was stalking, I moved often—my head, my free arm, my hand. I did this respectfully and calmly so they would know I was present. They understood and continued doing as they pleased.

I moved slightly into the sunshine so it would warm her while she rested in my hand. She seemed to like it. After the sun slipped behind a cloud, I began moving slowly around the yard. I checked feeder cups, moved a platform feeder with one free hand, and gently shook out the hanging feeder while she rested in the other. I wanted her to see the things I normally do in the yard when she watches from Trees. I did not want to focus entirely on her so she could relax. She rested and watched.

I moved to the other side of the yard. I let her listen to Birds above and receive their messages. Then I placed her in a large pot beneath some Flowers. She sat there and was able to hold her head upright, thankfully. She watched me slowly remove my outer layer and create a small nest from the shirt before placing her gently back inside it. She no longer squirmed or tried to fly. She allowed me to move her. Then I walked around the yard with her resting in the soft shirt nest. She seemed to prefer that to my hands. I returned to where I was and the cover of Plants and sat back on the ground. She jumped from the shirt nest onto the ground. She hopped again onto a branch that rocked beneath her as she worked hard to maintain balance. I placed my hand behind her and helped her stay upright. She finally regained her balance, turned, and faced me. She was perched on a branch on her own, and she was okay.

I went inside and watched from my office window as other House Finches came to the dense cover. They ate from the cup and rested in the cover of Plant. I saw her jump higher into the greenery. She was okay.

It takes some of them a while to recover from crashing into windows. So many suffocate and die when they only need a little help. Many break their necks or instantly damage their fragile lungs. Many are knocked unconscious and can recover if gently propped upright, without crowding and without sound, so they are not lying flat with the weight of their body restricting their airways. Glass does not exist in Nature, and the reflections of light and surroundings at different times of day are confusing. Loud human voices and intense focus overwhelm their hearts and stress levels, and many will not survive that intensity. They need calm, quiet, peaceful space—just as all of us need when we are unwell and struggling.

I went back outside and looked up just in time to see one male House Finch fly into the neighbor's Tree from those same Shrubs, with a young little female close behind who was still learning how to land properly. She landed on leaves while he landed securely on a branch. They were together in the sunlight. She was flying that slow, fledgling flight they use while learning and gaining strength. They sat in the sun, and I sensed it was her and that he was the male I had watched calling to her and comforting her throughout her recovery—her protective parent staying close and witnessing our actions.

The following day, a little fledgling female House Finch sat on the fence. She had tiny puffs of baby fluff peeking through the feathers on her head that looked like tiny horns. I went outside. She sat with me for several moments. We simply looked at one another for a while. I am certain it was her. A quiet moment that spoke volumes.

Birds do not like to be watched and certainly not stared at, but she and I looked into one another's eyes for about a minute. I felt compelled to do so, even though I normally never would. I was drawn in. I kept my distance out of respect. She then looked up toward Mountain Ash and flew into Tree. She was much stronger at flying than she had been while learning beside her father just one day earlier. She was well, and I wanted only for her to continue being well, healthy, and happy in the ways she needed and deserved.

In this experience with this sweet young life, our connection was bound in kindness and respect. I did my best for what she needed, not for what I wanted or for whatever else was waiting for me beyond that moment. My choices are mine alone. How I respond to what occurs around me creates my environment—rippling outward and flowing within. Our yard, the visiting lives, and those who call the space home were witness to it all—waiting and watching, creating more ripples beyond us that day. I have become known where I live through these experiences, most of which involve nothing more than basic consideration for others who share the space—our common ground. In this dire moment for this young life, our interactions moved through the air and Earth, through the breeze and Trees, and beyond. We create what we are, alone and together, throughout humanity and throughout Nature. I know I am not alone in offering this kind of kindness. I, alone, hold my key.

Fifteen
The Illusion of Difference

Living in Balance

We work to create space that is shared with life around our home in a balanced way. I work to keep balance within my focus and maintain it over time. Many of our homes have been in neighborhoods, and I remain mindful of this as well. Once I learned of my place in this equation, I began to think of our yards as small spaces to eat a little and rest—a place of safety.

We have lived in areas that can be very dry in the summer, without a naturally occurring water source nearby. These areas offered me, quite clearly, that all life among us is so very similar, even at a basic functioning level. Along with interconnectedness being among our basic needs, we need space, health, happiness, joy, and companionship to function properly and as we are intended. With all the needs we share, regardless of our outward differences, all of us especially need clean, healthy water.

Witnessing certain Wasps find the smallest drops of water that have produced little areas of softened Earth for their use is remarkable to me. Even beyond their incredible abilities and intelligence, they drink water and will seek sources in our yard. We have always had birdbaths for Birds. Birds need water like Plants and Animals of this Earth. So do Wasps. So do Honeybees.

In a short amount of time, Honeybee individuals find our current birdbath and relay this information to others in a way I may never fully understand. They visit and drink in steady intervals before leaving to return to what I can only assume are their beehives—three different directions for possibly three different hives. In a most recent location, Honeybees preferred to use a small potted Plant next to the bath that had a good amount of Moss growing in the soil; super sturdy and spongy. All Birds and Squirrels were drawn to this Plant, but none as much as Honeybees. So, I maintained the moisture.

I watched Honeybees for a little while one late afternoon in summer. This was their preferred time to visit, perhaps because it was the warmest time of day. Or maybe it was because they were thirsty. Or possibly because they needed more moisture, given the late summer and reduced Flowering and nectar sources. All this thinking my brain produced.

I watched one individual Honeybee arrive and drink through some Moss in a way that seemed very comfortable, and then little, mighty Honeybee flew up in circles and headed north. Two minutes later, another Honeybee arrived, drank, flew in circles again, and headed off to the north. They were like aircraft, perfectly timed.

As the weeks went on, I noticed other routes of Honeybees. Those heading to the north continued, but then those to the east and south began, too. All were spaced in what seemed like a coordinated effort that I lacked the ability to comprehend. What brilliant mindfulness.

This was not new to me, as I had witnessed Squirrels arrive to eat a few seeds in turns with one another, too. Some young males needed to be taught several times in Squirrel ways by older females—possible relatives and mothers. They took turns visiting, eating, and drinking, and it worked really well once a routine was followed. The same with Birds. Those Saturday mornings when I was an hour late putting out seeds, I caused so much disruption in their system and coordinated visits. Still, they figured it out among themselves for the day. To me, at first, it seemed like chaos and inconsiderate consumption, when really, I had disrupted what they had worked out quite well. Birds adjusted very quickly. What looked like one situation to me, based on our differences, was really very different in reality. Birds and Squirrels worked together despite me, time and time again.

One morning, I decided to move small potted Plant with Moss over to the other side of the porch, about five feet away. That afternoon, I watched Honeybees search and search, repeating circles and trying to find the pot that had been there. Using the Sun and location, they found nothing where there had been water for them the day before. I watched them arrive in timed increments, search over and over, and then leave. Each one was without water. I acted after a very short time.

The next Honeybee arrived, and I placed the little pot back in the original location. Bee landed and drank. They drank for a few minutes, as they would. As they were taking in water, I gently moved the pot to the new location. I did this for a good while for the others arriving, as I enjoyed some time in my favorite space of our home. Over and over, for those Honeybees who arrived, I would move the pot to them and, with them, place the pot in the new location. I was hoping to show them in a way that our differences could meet on common ground.

As each Honeybee left the pot and headed in their intended direction, they flew in new patterns around the pot and porch before heading out. By the next day, they knew the new location and found it immediately. I could only assume that each individual Honeybee went back to their hive and explained the new location for all to know. A little help—just a little support—can go such a long way regardless of who we might be with.

Not only do all of us need water in very real ways on our Earth, what looks different to us may not be different at all. Kindness, cooperation, support, and reactions to balance and imbalance are better understood at times when we stop thinking about what *might* be needed and pay attention instead to what we sense with *and from* an individual in the natural world. The same is true with people.

Different in Expression, Similar Needs

You might be familiar with the intelligence of Crows. Perhaps Crows are more willing to offer this side of their intelligent personalities to people because of who they are, their size, and their ability to remain safe in groups. Smaller Birds have more to worry about and may experience similar social and individual qualities expressed differently. Crows have been known to visit deceased members in turns, just as Elephants do. Other Songbirds and Doves experience grief and loss in solitude, from what I have witnessed. Always, that moment offered between two in honor of another's life and the depth of their loss is apparent—felt.

I have witnessed Raccoons checking on those just struck by a passing car, wanting to help and move their loved one. Many fear for safety, with the only option being to escape the scene as quickly as possible, while others hide in fear and, perhaps, trauma from the situation.

As such, life in Nature seems to understand the power of collective decisions better than we humans do.

These lives have ways that create a collective whole: a group mind. Our bodies—the Nature within each of us—operate this way, too, yet we still struggle in our current, advanced worlds to work in harmony with one another. Bees work so well together that, collectively, they resemble the functions within our human brains—this similarity between them and us has been more well-known lately. In the hive, Honeybees share information.

Food and water source knowledge is generally shared using the Sun as a locator. But it seems there is more.

It appears Bee intelligence is so supreme that we, as people, cannot know its breadth and depth. Bee intelligence exists on an individual level and as a working, functioning group. It would be unwise of us to conclude that this must be all there is when it comes to our similarities. Perhaps what we see here—and it is pretty amazing—is the proverbial tip of the iceberg. Like so many in Nature, decisions are made by the group, perhaps for the good of the whole.

Trees talk in their own way. They communicate in their own way with one another. They could also be sharing information through other means with other life. I can only believe other life throughout Nature is privy to the conversation and information. Perhaps they would share with us, too. Maybe they already are, and with all people. When we are sound within ourselves, through our senses, can we listen? Not in some poetic way; could we learn ways to support them, and by doing so, support ourselves? And in this support, support is reciprocated, as is the way of interconnectedness. Something within our natural brains and bodies shifts. We can feel it.

Ravens and Crows remember people and can recognize them regardless of the day, clothing worn, or hats covering faces. They know us. They remember one another, too, and through lasting relationships. Other life in Nature remembers as well. Is it so far-fetched to think this is the norm? Plants can "see" or sense without eyes. Similarities in DNA that Plants share with Animals and people also regulate responses to light. With Plants, I believe we become known as well.

We can appear to be very different, but within our physical selves and emotional selves, we are not as different as many of us would like to believe. Perhaps an embrace of acceptance—that we are alike even if we are not the same—and offering the world around us due regard instead of disregard might realign our personal wellness among greatness and, in turn, help us accept our own.

If an individual expresses themselves in a way that is unlike how you or I would express that same emotion, does this mean the individual in question is difficult to understand? Possibly. But people also express emotion very differently from one another. Maybe this is a poor example given our inability to communicate effectively at times, and often. People tend to be less reliable, given our love affair with talking and abusing our gift of language, leaving what we express as not always an accurate representation of what we are feeling or experiencing. Many of us have grown to utilize a poker face so well that we forget what we are even feeling, perhaps removed from our connection to it as a form of protection that has served us so well for so long.

We believe life in Nature is difficult to read because many of these lives lack facial expressions. Yet those who know particular groups of life well can distinguish expressions quite easily and readily. To us, we might not see the changes or expressions; they remain even if lost on us. Emotion and communication still exist, even if we do not approve or believe.

As beings with emotions, we are often moved to compassion. Again, we cannot claim to be the keepers of compassion, for this strength exists throughout life in Nature, among bonds and friendship between Animals within the same species, and life across species. Elephants, Whales, Cows, Pigs, and Doves have deep friendships and relationships within communities. Seahorse pairs spend time together before their days begin in an embrace. Life in Nature helps one another and grieves in loss. They remember over time and are joyous with greetings upon return. It is not so crazy to think this is common to all life. We could be more like their example.

Rooted in Wisdom

Houseplants are a wonderful source and reminder of interconnected inspiration. I have always found, and therefore believe, that Houseplants live better and healthier when not alone, but with another Houseplant in the room—two independent pots instead of only one pot. They need one another. Plants living with me are healthier and appear happier, with more vibrant color, when they are closer to another and within a certain distance in the same room.

I touch them gently before any movement, so they are alert and aware that they will be moved or that I am about to water them. They thrive on kind, gentle contact just like a breeze, rain droplets, small Insects, or the brush of a passing individual. I handle them carefully, understanding their ability to sense me and my movements and the fact that they are bound to their positions and pots. I have great affection for these indoor Plants and do think of their needs, as best I can in my limited ways to understand and learn from them.

During times of soft breeze and gentle weather, I open windows so they can sense in the winds what other life in Nature relies on to know and be. They seem fuller with life after these moments. These experiences, among many others, do not take much from me and offer me much more than I could ever give them. They make my house a home. They are beautiful lives that share a home with me. They are present and radiate presence to me and with me.

I am aware when they need water by simply stopping and entering that present state with them. In this knowing, this silence, we enter a common area for our communication to flow between us—interconnectedly. In this state, I feel refreshed, balanced, calm, and happy. I feel love. I feel connected and embrace this welcoming connection with any indoor Plant whose path I cross; gently asking for permission, without spoken words, to kindly, gently, and softly extend my hand as a form of contact in acknowledgment of all that they are to me and to life.

Those of us held in place by Earth, incapable of moving about and exploring, know of their surroundings, help one another, and are aware of the intent of others—including humans. Plants use numerous sources of information that are communicated through the air and soil, from and to one another, and from other life forms to understand their worlds. Plants, including Trees, are aware. Different, yes, but with important similarities and their own exceptional strengths.

In our kindness of offering consideration to these lives all around us and within our homes and offices, could we extend to them an admiration of their essential value and trust in what they are able to do and be, even in our limited information and understanding? Would indifference be replaced with respect? Perhaps then we might be able to see our own gifts a little more clearly, given our ability to shed light on interconnectedness with these lives and the inherent worth of those around us, throughout the air carried on the breeze and radiating with reverberating life under our feet.

Perhaps it is easier to belittle them, given where we place Plants within our worlds. This does not make it right. In understanding them, and in understanding that we do not understand them completely—given who they are, who we are, and how they want to live, too—maybe we could take their worlds into account.

When we see them for who they are—not as objects, or disposable things, because they are not things but individuals—we might rethink their care and happiness. This would only help our own care and happiness. Perhaps we might find that we need more balance in our own lives and that we need trusting relationships with them more than they need with us. We need to be trustworthy. All in all, we do get what we give within our bodies and in our minds. A clear heart beats more easily.

We have much to repair in our relationships with life in Nature, in the wild, in natural systems, in our communities, and around our homes. We can come to a greater understanding and treatment, and find new ways to be with each other. The treatment comes back to us—to our bodies, brains, and minds. Our happiness and health, beyond physical wellness, are altered, too, because positivity does create positivity just as negativity creates negativity. Positivity does even more so within our worlds and within our Nature. Making the steps to change matters with our intent.

When we look in the mirror, it is time to see who we really are and whether we are in need of healing in these relationships. Because these lives are not that different from ours, no matter how much we try to rationalize our treatment and outlook.

Is Difference Only in Reflection?

Lives throughout the natural world have needs just like we do as people. Their fulfillment of those needs might look different from ours and perhaps be different, given who they are and who we are, but we are the same in that needs exist.

If all life fulfilled needs in the same way, the world would have collapsed a long time ago. The differences are vital to our existence. If everyone required nectar as a main food source, balance would not last. Diversity of life is not for nothing. We need diversity to survive—all of us.

We are beautifully and uniquely different for a reason, because it is the only way that works in the long run and for the long haul—a possible lesson for those of us feeling lost in an unnatural need to be like others. Getting along does not mean being the same. Getting along holds space for respect. In this, there is no home for hate and awfulness. Perhaps Nature might offer us our own inspiration about acceptance.

Life in Nature still operates without the burden of the endless options that we carry, and perhaps this is a small part of why we feel at peace among their ways. Perhaps it is why we feel the need to "escape" in order to find ourselves among these lives. Our minds ease with them.

As people living in today's world, we are flooded with so many choices in such short amounts of time that we tend to miss out on enjoying the moment. To be present is naturally fundamental to our health. Too many choices strip us of finding value in what is special because everything is available unlimitedly and without much regard for being enough. As such, we often do not know what we are thinking or feeling to clearly see what we fundamentally need. By knowing Nature without judgment, we might learn from their ways, given our struggles. Within each one of us lives the flicker of memory from being closer in understanding with Nature.

Plants have a need to be understood and a need to care for those around them. They desire to be supportive. They respond within their bodies when misted with water. Could they also be reacting to us even without our knowing? Perhaps they feel in ways we have yet to understand because of Plant awareness.

Our beautiful Earth is the natural trigger for inspiration, perhaps to give us hope.

I have witnessed Birds on numerous occasions face the setting sun as if taking time to pause in observance. These are moments of beautiful calm, when they appear to be more relaxed. It is as if moments like this are treasured. Maybe they are enjoying an understood time of day to witness beauty, just like we do when we allow ourselves. Perhaps they are learning, based on past experience and memory, what they were taught and shown from the light and colors and information in the breeze of what is yet to come.

If we can find moments beautiful, why would other life in Nature be any different, even if their idea of beauty may or may not be different from ours? Primates in the wild have been found to leave stones in certain places, offering pause as if in reflection. Could inspiration, reflection, and beauty mean something more to them as these things do for us? We do not need to know in order to appreciate the fact that so many commonalities matter to all life.

Rediscovering Our Inherent Worth

When we look in the mirror and see ourselves for who we really are—without facades, without artificial egos—we can truly begin to see. In seeing our honest selves, we might feel lost. This comes from standing in a new perspective. Only then, guidance comes in gentleness and quiet, without words or toxic emotions. We can feel the difference when we see ourselves.

In these moments, when our focus is on appreciation and love, we find common ground with the lives around us in the natural world. It is here where we are accepted only for who we are—in our fears, insecurities, troubles and worries, strengths and joy, and dreams. Nature sees no false fronts.

When we look in the mirror that is Nature, we are seen. This is our ultimate support in the interior of our Nature within. We are supported to follow our passions and thrive in our lives when we radiate positivity with all of life on Earth. This is the nature of Nature.

When we rediscover our inherent worth as a result of our own acceptance, we are honoring the world around us. In this place, we allow and extend dignity to life in Nature. When this is extended to the lives throughout the natural world, our worth is mirrored back.

We do not need to look far—our yards, homes, and neighborhoods are full of life seeking to know us in authentic relationships. We are automatically accepted for who we are, and when we commit to offering the same in return, life is exquisite.

Change is also wonderful. By understanding our own uniqueness in life and learning from Nature instead of always trying to improve Nature on our terms and without their consideration, we naturally evolve and adapt. Evolving does not mean disposing of ways that have worked well for the majority of life. Adapting does not mean resisting change, for that will be our greatest downfall.

Life in Nature consists of masters at adapting to change; we are not doing as well. It is time to look to Nature for guidance and then hear what Nature needs to say, in their own way, without projecting our wishes. All of our differences make up this world, and that is our collective strength.

These lives are worth our time and care for their ways. Life in Nature knows how to stick together. We would assign the human term "loyalty."

Friendship and support exist all around us. Just as we miss those we are separated from or have lost, life in Nature feels, too. They might appear fine because of our differences, but the struggles are the same. Again, we are more alike than most of us realize. These lives around us deserve that recognition.

We would be lost and suffering greatly without them. We need these lives more than they need us. We exist because of them. We have homes, grow, heal, and live because of all life in Nature. They have one another. We have placed ourselves on a concrete and digital island, and we are not doing all that well. Perhaps we might have a stronger sense of self through our relationships with life in Nature.

The next time you walk past fragrant Flower or amazing Tree, understand that they are aware of you. Show kindness to Plants and Trees in your yard, on your street, in your home, and around your neighborhood. Offer them space and what they need. Connect with them. They are aware of kindness just as they are aware of aggression. Knowing this, do you see worth?

Life in Nature feels good when they feel good, terrified when terrified, and content when content. These lives seem to have a far better grasp on the importance of one another than we may. We have forgotten this or have grown and lived without awareness of it. But our bodies are aware of this loss—our Nature within.

Just as bodily functions happen without our full knowledge, neglecting our interconnectedness based on baseless fears, superiority, masked insecurities, and keeping these lives at a distance as "it," leaves us missing something in our bodies, hearts, minds, and brains, too.

There is much for us to gain from life in Nature through healthy relationships...through respectful relationships. But can we be trusted? Do we see the worth in being trusted?

We are taught to survive and succeed one way with people, but those same instructions do not work with individuals in Nature. When we sit quietly in our yard, honoring that we are guests in their world regardless of whether we "own" the land outright—without speaking and with calm and peace in our hearts—when we disappear into the world instead of standing out, dominating, or being a force to reckon with, we can simply be. *Us, together.*

And Nature, for the most part—the silent observer to us—notices and responds. Observing for a short time as a guest, and without the need to be the focus but simply one aspect of this much larger world of our small, even tiny yard, we see evidence that trust is prevalent throughout Nature. Interconnectedness flows to us and through us.

Crow Next Town Over

My husband and I had just come out of a market when I heard Crows on top of an adjacent building. There were two of them. They appeared to be a younger one and an adult. The adult left the younger one on the roof about thirty feet away and flew toward me low—about three feet above my head. My husband witnessed this, too.

I stopped what I was doing—handing something to my husband by the car—and was mesmerized by Crow. I had a baseball hat on, and, as they flew over my head, dropping down as if to get closer, they turned their head as if trying to look under my hat, closely focused on my eyes. Crow then rose higher into the sky and flew off with purpose.

When I stopped what I was doing because of the attraction I felt with Crow, my husband could not reach what I was handing him because I was not paying any attention, and he lost control of our shopping cart. He did see Crow drop down toward me and approach. I sensed that this Crow remembered me. I knew one Crow family very well at our previous home, which was about eight miles away. There was a day when I left unsalted walnut halves for two Crows outside of this market. Perhaps this Crow was one of the two, or from the family I knew. I will never know for sure, and that is not the point.

What I do know is that this Crow was trying to see my face. This Crow was searching for my eyes. We had only brief moments together, yet they created a lasting impact on our relationship. I became known to this Crow—a connection that remains amazing and filled with feeling. Even in our differences, our relationship remains healthy and strong. There is a knowing between us now, one that has carried over through time and space in different geographic locations. Trust and inherent worth greet us through how we hold difference within our hearts—our Nature within.

Part Three

Us: Who We Are Within

Sixteen

Sound Motive, Honest Intentions

When Change Becomes a Doorway

Change is one of life's most misunderstood gifts. It can frighten us with uncertainty, yet it also offers growth, renewal, and possibility.

Change can help pull us out of a slump. When our motives and intent are clear, change becomes a doorway rather than a wall.

For many of us, problems arise when we adopt ways of living that "make our lives easier" but cloud us from seeing everything else around us. While some of these advancements have done wonders, we are largely bombarded with noise and technology junk food. Our brains, bodies, and ability to stay sharp and grounded with others in a very real way—a natural way— are now largely hindered. Hiding behind screens and names, the motive and intent of our actions can become unclear and even abused. This is not a good place to be, for it certainly begins with the personal relationship we have within ourselves. Our actions directly reflect our happiness.

What is wonderful about change is that we *can* do it. We can adapt because we have been designed to do so and have done it in the past.

Change can be the best thing for us when viewed through the correct lens. The unknown can be daunting, scary, and intimidating. But it can also be wonderfully positive, exciting, and rewarding. It is all in how we look at it. With this perspective, our motive and intent for change itself become clear, too.

Since trust is missing when we act in ways that are untrustworthy, and life in Nature watches for this, we would be wise to raise those in our lives with the value and integrity of trust. For if we cannot trust those in our lives, and if we do not honor our own self-trust, life in Nature is less willing to hear our side and will do so with great caution. And rightfully so. We are never perfect, and we can't want perfection, since this is not real. But our actions and reasons behind them matter. That is, what is in our hearts matters most.

I have been watched, sometimes for an entire year, for my credibility to become known accurately. During a time of rebuilding my sense of trust with others, I ultimately shone light on what I was missing in self-trust. So I worked on what I had control over—myself. In return, great trust was afforded to me, placed in my care, and sheltered in my yard.

We have to start behaving in ways that are trustworthy—and that includes what is best for life in Nature. Life throughout the natural world is most important. *We would not be here without them*—this is important to reiterate. Because of them, we are nourished and grow, heal from injuries and surgeries, and have places to live. We can thrive *with* them. They would be better off without us as our sacred Earth shudders in attempts to restore balance for the good of the whole. But we need to be open to a trusting relationship with life in Nature, where we also take into account their right to live as they need.

Sharing Intelligence

Nature is often a "thing" we go to in far-off lands to find ourselves while disrupting their worlds. Void of this trust, our energy becomes one of isolation, deprived of important experiences we would naturally know and understand through our built-in receptors. This can leave a void—an emptiness.

Moments in life happen—more often than not—without words, while being present, through an experience with another in areas and communities of *home*. At a cellular level, we become part of a greater whole through real-life, in-person interactions, communication, and experiences. These are relationships with individuals in our yards, with Plants, and with those who visit these Plants—within our neighborhoods and communities; among individual people learning from individuals in Nature, learning who they are, what they need, and honoring their boundaries and space to be.

Through reciprocation, we all grow in our own ways. Humanity can remain present and without ego, allowing trusting relationships to develop, with real information passed and received.

Radiating from Within

As people in our dominant world of language, with the human-centered blinders we often wear, we can manipulate our behavior through deception to benefit our intent or plan. We can also be fooled by others who are master manipulators. People lie, cheat, and express what they think others need to hear.

In moments of manipulation, we can fall prey to the awfulness of some. This can be especially true when we are looking for support and understanding. Boundaries are essential because of this.

Our motives and intent extend as our "plan," from which our behaviors become known by Nature. Life in Nature has grown more and more aware of our ways and what we are doing to Earth. We generally enter their worlds without much regard for what they need—their space, their desire to be left alone so that they can focus on living. Their communities are shrinking, polluted, or overrun by more and more people.

Creating a space of trust does not mean having another trust us enough for daily interaction and contact. Creating a space of trust means offering clear, *consistent* intent from within—our inner self represented by our *daily actions* that also honor another's needs, space, and desire to thrive in balance.

Let's leave Nature to Nature. Can we step out of ourselves enough to assess the situation, offer quiet respect, and leave them alone? Can we act on what they need, too?

For far too long, we have focused on only what we need—to Nature's demise. It is time to place them as a priority instead of our flippant, harmful, and devastating interests. For without them, we would not be here. Choking them out will continue to impact us physically, mentally, emotionally, and spiritually.

Starting around our homes, neighborhoods, and communities, accept life in Nature with care. The diversity and beauty are apparent if we take notice.

A community cannot thrive unless we *all* have what we need—this includes space and freedom, health and safety—*including* life throughout Nature. Who we are, and our actions, impact all of Us. Take ownership and pride with, not against, life closest to home. Be honored to give them their space so that we can all be together.

Experiencing Gratitude

Gratitude is something I have experienced with Insects, Birds, and Mammals in the wonderful yards where I've had the privilege to live. They've taught me an incredible amount about gratitude, the need for gratitude, what holding it within creates, and what expressing it in various ways means beyond my own existence.

Our motives are known—often by people and always by the lives of Nature. When they arise from care, respect, and gratitude, trust deepens and relationships thrive. From the individuals in Nature around my home, I have learned how common gratitude exists among those who trust.

Expressing gratitude is wordless from the natural word. Often, it is through long expressions coming from moments when Chickadee will pause what they are doing, move in close to me, safely, and hold contact with my eyes—motionless. This is not me focusing on them, since many find it uncomfortable. These are moments when *they* stare through me to my heart, and I feel pulled in to *see* them. After seconds that feel like minutes, they leave, and something within me is felt, sensed, and interconnectedness is acknowledged. I am then altered in a wonderful way.

I feel more gratitude. I hold gratitude for the seeds that helped many survive in overdeveloped areas, including my home, during harsh winters. I become incredibly grateful that lives in Nature trust in me by knowing I mean no harm, and that I only want them to live as they deserve, respectfully. As I act and behave as best as I can, knowing I can never do everything fully right by these lives, our motivations and intentions are known by life.

I've had very special moments with individuals wanting nothing from me. We share a focused connection, and through those moments together, that's all that is sensed and felt. These moments taught me to incorporate gratitude every day, in quiet moments of meditation. These expressions show me how to acknowledge all that I am grateful for in my life as

I go about my day. I now do this every morning. I am so very fortunate for the brilliance of these lives in my life. In moments and over time, these are experiences and information that have changed me forever.

Seventeen

We Meet at the Bridge of Inherent Worth

Embracing Nature Beyond Thoughts and Words

We can learn to "hear" Nature in ways beyond our words and thoughts.

Once we become more in touch with our inner world, this allows our suppressed or unrecognized senses to open. From this place, we can inhale deeply in kindness with Nature, and we can allow life in Nature to be who they are.

Honoring boundaries and leaving behind any sort of dismissal, we may learn from these intelligent individuals and create supportive relationships simply by being together. We then gracefully and acceptingly feel emotion. We become bathed in these enlivened experiences filtering through our senses, and our thoughts are then filled with inspiration and encouragement.

With Nature, we breathe life into our conscious minds, our bodies, and our cells like a calming breath of life within.

It is time for us to embrace Nature as they once were before we forced them away. We cannot flourish in our daily lives without the support of our surroundings.

Seeing Through Nature's Eyes

Nature is communicating all the time. We must begin to learn to listen. We can do so by dropping what we might understand as communication and allow Nature to show us what they need. And the trust they need from us—which will only happen if we become truly and honorably trustworthy—includes leaving them alone and letting them be. We need this within ourselves and our communities—before life in Nature is no longer there to be heard. Because once their brilliance is gone, our struggles will grow, widen, and deepen—personally and beyond.

If we begin to see Nature through Nature's eyes, we see ourselves from a place of honesty and kindness. We are accepted for who we are if we provide the space for another in trust. We begin to see that we are not so different from Nature. If we are not so different, might we learn from one another? And if we could learn from one another easily in our daily lives, would we not develop trust for one another? And once sound, true trust is established and held dear, would deep respect begin to take root? And if we operated in a world where we felt respect and trust with Nature, would we begin to feel a deep sense of community with Nature? And once we felt this deep sense of community with Nature, would we not begin to care deeply for our supportive community? And if we cared deeply for this supportive community...would we not want to protect them—Nature, these lives that *are* Nature—and ensure Nature's safety and balanced survival so that life could flourish?

When Nature is balanced and supported, we can become part of that environment, too. We begin to *feel* for Nature. And once again, we are most likely to protect those we care for deeply.

When we learn to understand common ground and how others relate, we learn and grow ourselves. Our level of relating and understanding deepens through Nature's way of relating to the world, and our respect grows. Through these mutually supportive relationships, we begin to see similarities between our world and Nature's as we learn from Nature, for we can see that we need one another.

We are interconnected. We are interconnected with one another. We are interconnected within our bodies, our brains, and our minds. Our way of being extends far beyond our physical place and seeks this balance with all life in Nature, too. This is our support. This is where our strength, our inspiration, and our counterbalance of health and happiness live. It is not a choice. It is unspoken. It simply is. Nature thrives through and utilizes interconnectedness together. We have forgotten this part of ourselves and, like a dry lake, it waits within us—parched and cracking for the steady, healing raindrops from our intelligent relationships with life in Nature.

As we walk along life's path, we find ourselves at the crossroads with these individual lives and communities of Nature. Here, we pause at inherent worth.

In Honoring Boundaries, We Recognize Inherent Worth

I only recently realized how vital boundaries are to creating so many good experiences within our own lives and with life in Nature—the role of boundaries in regard to trust and how boundaries support another's worth.

Boundaries are the bones of our being. They uphold our personal well-being. They create the structure and strength of who we are and affirm our inherent worth.

We are not taught much about the significant role that boundaries play in healthy relationships. To consider another is to allow for, expect, and respect their boundaries, as well as our own. Honoring our own boundaries frees us to honor the boundaries of others, including the lives throughout the natural world.

Boundaries are essential in supporting and being considerate of the fundamental, inherent worth of another.

The Potency of Worth

What is worth, exactly? Do we feel it, see it, sense it, and know it?

When we recognize worth and place value, we see importance. Perhaps we can assign worth to many different things, situations, and even attitudes. But what about individuals? Where does worth come from, and what does it mean for us?

Do we see worth in others? Do we see worth within ourselves? What can *worth* do for us when we *feel* it?

One could argue that self-worth is the most important form of worth when healthy and balanced. Yet, so many of us fall short in believing in—or even knowing—our own worth. Perhaps in our personal challenges, we also fail to see it in others.

Knowing worth is intimately tied to our interconnectedness with life in Nature. Each day, as we notice and believe in the inherent worth of life in Nature, we extend our trusting boundaries to the lives that still *do* embrace trust. Interconnectedness is the vein that carries this flow of understanding—this *knowing*. This is the fundamental progression: from worth, to trust, to respect.

Life in Nature has worth within their own right. We are the ones who must mature to know and appreciate this fact.

We Meet at Inherent Worth

With interconnectedness comes an embrace—an embedded understanding of another's inherent worth, regardless of the life form that individual may take. In this state, our physical, mental, emotional, and spiritual selves respond positively to life in the natural world, innately and effortlessly. It is where we go from here that becomes our choice

Our inner selves are incomplete without the support of life in Nature. When we attempt to sever this connection from within ourselves, we only hurt ourselves. The ripples extend outward from there. Our relationships with the individuals who live near our homes, neighborhoods, and communities—and, in turn, reverberate throughout all of Nature—link our worth to theirs.

The foundation of trust is the underlying current that supersedes any words that are spoken. In the fertile soil of trust, we come to know that the lives we meet, whether for moments or for lifetimes, possess inherent value in their own right. In this recognition, we begin to see the first light of respect rising on the horizon.

Bee Fly Resting

It is difficult to know another's worth fully when we have not focused on the foundation of any relationship—*trust*.

In the suburban and semi-suburban areas where I have lived, I've worked to create yards that offer boundaries and space for the lives who choose to share these places with me. Over time, these spaces have become communities of individuals who have taught me to recognize worth through trust. In responding to me, they continue to teach me more about what it means to be trustworthy.

Trust comes from a deeper knowing and the relationship between individuals. Becoming trustworthy ourselves is the catalyst for change. I have had to learn to change.

First, I needed to trust myself in my experiences. Since my experiences link me to these individuals through our senses, I have felt when I needed to better honor my own self-worth, my voice, and my worthiness of love so that I could learn these same ways from these lives. And I had to start trusting outwardly what I had always known within.

We were out on a small hike in a very beautiful, arid area near our home. So fast, a Bee Fly came in close and hovered in front of me. I trusted this special little one. More importantly, I trusted myself. I knew to stop and pay attention to the sense that I needed to place my hand palm-up in front of me because they were in need. So I did. And so did this little one, in their own way: They rested for several moments on my finger as they cleaned their face and legs.

Beyond their highly beneficial life and their flight, which is so enjoyable to watch, I could get lost in their movements. But, in this moment, something happened: I began trusting myself by honoring their worth. My husband witnessed this and even took a photo since his camera was ready for images of Birds and other beautiful sights. This was a very simple act for what many may consider a very simple life. They are more to me.

With Bee Fly, I found the stepping stone to my voice through knowing with senses and this form of intelligence in a way that has guided me forward profoundly ever since.

This moment of knowing one's worth alongside these lives and natural areas and communities changed me forever. Why—or better, how? On this very hot day, our mutual trust was acknowledged because I was able to honor my own worth within their worth. What I sensed was validated once I trusted myself enough to act on my inner knowing, and when I allowed the Nature within me to respond to the Nature around me. Interconnection is in action within silence.

Eighteen
Earth's Living Intelligence

Intelligence: A Brilliant Force

More and more, we rely on technology to recall, remember, learn, and think for us. The farther we remove ourselves from life in Nature—and from our own intelligence—the more our world, within and together, reacts negatively. We must appreciate the value of both. Balance is required here.

Each individual in the natural world is a brilliant force, just as each one of us is. Accepting this truth honestly begins the opening of our own intelligence and helps us find our voice in this world. Strength and power come from within and from honing our gifts, just as life in Nature does.

Intelligent brilliance is occurring beneath our feet. I remember learning as a child that it takes many, many years for even an inch of soil to be created through the natural breakdown of matter. Every time I see soil dug up and dumped in large amounts, I still think of this. How special this process is—for us to have soil, something all life needs and eventually becomes. All of our food comes from Earth—from soil. Healthy soil, all the communication flowing through it, and all the life it must support—we might be wise to treat it with care and respect. All life grows and feeds from soil. Soil matters. Soil flows into our waters; interconnectedness within these silent interactions builds what we—all life—need and become on this very special planet Earth.

I once learned of zookeepers noticing the signaling of Elephants at a certain time each year and eventually realizing, after years of observation, that a circus was in town miles away. Captive Elephants were believed to be calling back to those they would never know or meet, who would spend each day forced to serve human entertainment. Zoos, and those deeply committed to helping wildlife, work hard to care for and support many species so they can remain on our Earth and live well in their lives, including Elephants. It's a double-edged sword.

Could our needs, at the expense of the health of others, be doing something far worse within our inner world? What we see, experience, feel, and suppress pulses through our brains and bodies, creating a consciousness that does not fully account for kindness but instead seeks brief moments of escape through entertainment. A sickness is felt within our gut, telling us—beckoning us—to understand that something about this is not right.

How much these lives must adapt to challenging environments—droughts, floods, currents or lack of currents, climates, temperatures, and fluctuations—is mind-boggling. The current rate of change is testing all of us. It does not need to be this way. It is not meant to be this way. We could be just as positive a force on this Earth. Finding our positivity and acting on inspiration receives the most attention from the natural world, and this is what is needed. Interconnectedness is a state of mind; a natural condition of our body. Look within to change our outer world.

Silent Knowing

There is empathy, and then there is *knowing*—the deepest roots that seem to interconnect with all roots. Consciousness surges among all intelligence on this Earth.

Nonverbal communication generally refers to the behavioral elements of human-to-human messages beyond spoken words. One's appearance, posture, and facial expressions send messages to others and provide additional cues to meaning. Much of our emotional communication also comes from nonverbal sources. Our mental state and emotional position together create an energy within us. As these states affect our biology and physical condition, this energy is created and radiates from us. Nature knows us through these ways, combined with the active energy in the world—our actions. Knowing extends beyond verbal language, encompassing our mental, emotional, and physical selves.

We feel what another feels. For those of us who excel in this ability, there is more to it and more to how it connects to life throughout Nature. For those of us biologically wired this way, becoming well-versed in it may take considerable time to understand.

It takes self-trust to recognize, use, and become familiar with this *knowing* in a grounded, kind, and steady way. We cannot force it to happen; it is present when it is present. This strength is an invaluable intelligence within our world, with others, and with life in Nature. Any misuse of this gift or false presentation can quickly diminish it, leaving us unsettled, unkind, or out of alignment—and very perceptible to the life around us. Honor and integrity are necessary.

Sea Life

Life in Earth's waters and seas lives with one another in ways we will never fully know. But they matter to one another in ways that support one another, along with so much other life, through interconnectedness.

We can recognize both the communication taking place and the interconnectedness of life in the sea. Coral decide where to live. Once they do, they cannot change their minds and leave. Coral hear and respond to Fish communication. They are aware of what is going on around them.

Dolphins have been held captive and confined in small spaces as displays for people. Just as we would, they can lose their mental health and go mad, swimming endlessly in very small circles within very small spaces.

Fish are amazing and highly intelligent. This should come as no surprise, given all they must do, possess, and become in order to breathe—live—let alone reproduce. They remember, and they must remember to survive. We would do well to learn a great deal from them. Fish can recognize and recall people when housed in tanks, and they understand how to work together when left to live in their home waters. They are not mindless or without feelings. They simply express themselves differently. They will never mirror our own physical ways, but that does not mean they do not feel or that they are not highly intelligent.

Fish experience a rich range of emotions, many of them much like our own and deeply influenced by their senses. They feel joy, and they do feel pain. Looking and living differently does not equate to lower intelligence or an absence of emotion. What might our treatment of them reveal about us, and can we begin to approach both their needs and our own differently? Might we feel better in the process?

Changing our attitudes toward these lives in Nature so they are valued for the individuals they are—including their intelligence—may allow us, in our own ways, to embrace them without breaking trust. Let them thrive within their own boundaries, just as we need. Help Fish thrive in their worlds. Healthy water—the blood of Earth—matters greatly for everyone and all life on this Earth, yet it is so often taken advantage of and taken for granted. Water is essential to all life on this planet. *Clean water*—without it, there is no Us.

The Understanding of Squirrel, Hawk, and Goldfinch

There is a male Squirrel who I think is Stripe. This is a new relationship, so I am not yet fully familiar with this individual. He is very light, happy, curious, and likes to spend time in the yard with me, watching me while casually following me around. He seems comfortable near me and often turns his back while he does his thing, and I do mine. Sometimes he simply likes to be with me. I welcome his company and do nothing to encourage it except avoid speaking around him so I do not startle him. I move slowly and respectfully. This friend likes hopping around the yard and running along the stones on the ground.

I was in the yard just before a cool autumn evening. Female Goldfinch was eating thistle seed from the feeder, and young male Cooper's Hawk was in the yard. Goldfinch was very hungry, and I think Birds are coming down from the mountains when snow falls in the higher elevations. She would not leave the feeder as I stood next to her waiting. I wanted to take the feeder in and store it as I do each night. I waited patiently.

Hawk had just flown into one of our windows—a sign to me that he was young based on experience. He had turned at the last second, so it was not too bad an experience for him. I have learned to align curtains and panels of sheer material with pretty designs that align vertically to create a sense of boundary so life can see the window as an object and not attempt to fly through it as vacant space. Sheers let beautiful light in because I need that, too. It helps more often than not. Hawk was okay and was back on top of the feeder post. I went outside because Goldfinch was still at the feeder and now frozen in Hawk's presence. I have witnessed Birds freeze and enter a trance-like state when in danger. Hawks focus greatly on movement. They seem to zero in on motion. Goldfinch was frozen, and Hawk was now about four or five feet from her.

When I establish boundaries clearly within myself and while I am in the yard, through thoughts alone and gentle reminders, those boundaries are mostly honored. To this day, I am surprised by this. Nature and these lives are so brilliant that I am often left in awe. I try not to question or overthink anything. I choose mostly to sense. This time, Hawk would not fly away. I felt part of this experience in a way I choose not to assign words to. Words do not seem to fit well. There was an understanding taking place. That is what I *know*.

I walked slowly and respectfully right by Hawk, and he stayed—three feet from me. They often act as if they do not see me when they clearly can (another common occurrence not tied to any particular house or area). I stood between Hawk and female Goldfinch. I put my hand up to block her view of him, but she was not looking. At that point, I did something I would not normally do: I pushed lightly on her chest for a small, delicate Bird, but she would not move, so I pushed a little harder. She snapped out of it—out of this trance—and flew off balance to the rear of small Tree. I stayed because she was still nearby, and so was Hawk.

Then Stripe came out from some tall Trees during all of this and sprightly hopped over to me. He climbed onto the fence while Hawk sat on a post again. Stripe stood between Hawk and me. He turned and faced Hawk, and Hawk lowered his head, puffed himself out, and stared at Stripe. Stripe stood his ground, pointed his nose back, and stared in return.

These interactions can be felt quite easily. What transpired, however, belonged to their own understanding. They had some sort of quiet conversation and interaction. Hawk then flew back behind the growth but remained very close. Stripe sniffed the air toward Hawk's direction. I knew the little female Goldfinch was still around. This Hawk would not leave with my snapping fingers or banging weather domes from birdfeeders together—again, something I never do—but everyone stayed put. My brain was working too much. Goldfinch, my Squirrel friend, and Hawk remained. This was the extent of how annoying I was being, given that I do not like being knowingly disruptive. Since there was some sort of understanding taking place, I soon let go of my need to understand it completely and allowed myself to know that it was time to leave. They know far more than I do.

I walked back to the house and quietly thanked Stripe for his help as he walked with me along the top of the fence, bounding in what I would describe as proudly. My husband witnessed all of this from inside through the large window. Afterward, he told me that Stripe walked back with me as though we were a team. In the yard, the feeling was more about community.

After some time inside, I decided to go back out because little Goldfinch had returned to eating, and Hawk was again on the fence. She was no longer frozen, just eating. He was about eight feet from her. Her back was turned toward him. He was not interested in her. It was almost dark, and she continued eating and eating. I stood between them and stayed with her while she ate. Hawk did not acknowledge me much, but when he did, he stared directly into my eyes. I tried to convey that he was welcome, along with everyone else, but that I wished for this to remain a safe place for all.

Hawk eventually flew away. I am unsure whether he did not see me or simply did not care that I was in his way. Perhaps he was establishing a relationship with me through his own communication. He was not looking at me but flew directly toward me. I had to raise my hand to block him, and he saw it. I heard the air shift through his redirected wings as he darted up over my head and around me. Goldfinch continued eating.

Hawk disappeared from sight. I stayed about a foot from this hungry little Goldfinch until she finished. I have never seen such a hungry Bird. She was desperate when she finally found food. I continued looking for Hawk while standing near her until I heard her fly away. She went to rest for the evening. She seemed healthy, perhaps just a little tired. Maybe she was migrating or finding hardship with the decline of natural food sources and proper habitat. Not all Birds are the same, eat the same, or live within the same habitat. Not all migrate, either.

It was an amazing experience for me in understanding how they know one another. And how Stripe, the Squirrel, helped me—almost translating for me. The three of them understood one another through ways of communication I will never fully know. Yet they understood me. They know so much more than I ever will. My thoughts rationalize everything away. Only when I stop long enough to sense and *know* do I receive fleeting moments of insight.

The Return of Consideration

Whales protect other wildlife from danger and attacks, including people. This is yet another example of how consideration for one life by another is prevalent throughout the natural world. Consideration resides within us, but many are out of touch with this important part of the self. This capacity for care and responsiveness is present across all individuals throughout Nature. As much as we may try to claim it as uniquely human, doing so contradicts what is natural.

It is natural to feel and take others into account. But we have, for the most part, lost touch as a species, and the mirror of Nature reflects this gaping hole. Emotional intelligence thrives all around us—layers and layers of information, understood through vast senses, coloring our worlds and the worlds of individuals in Nature. When we function together as a supportive unit, we thrive. Being considerate of others is paramount.

So many of us are not feeling well at all. We have turned our backs even among our own people. For life in Nature, consideration, along with the emotion and intelligence involved, is necessary for interconnectedness. Working together in mindfulness of others allows communities of life to operate fully. Needs are met; consideration is restored. There is no logic in feeling superior. Empaths and those who are highly emotionally intelligent are straining to find camaraderie. Look to life in the natural world; we are at home here.

Do we appreciate our own brilliance? Perhaps if we take a few moments to sit quietly and respectfully with life in Nature in our yards, on our porches, or on our streets—with an individual Bird, Plant, or Insect—they might welcome us into their moment within the day and shed some light on what is possible. Remaining mindful of the impact of our presence and allowing another space to exist in calmness is *progress*.

When we can remain present in our bodies during moments that push us toward emotions that are not good for us, we learn to appreciate the strength in calm. Calmness is strength. Calmness is a balanced state within that can be felt by the world around us. Here, we learn to appreciate ourselves and our weaknesses—for we all have them—and take steps toward our inner self, our more aligned expression of who we are. In calmness, we no longer need warring shields to protect us. What protects us, and who we are, is our sense of self.

We matter, and we hold intelligence. We contribute in our own unique way to this world. In balance and calm, we can begin to truly hear another's voice. We notice, in flashes and moments, the intricate and vast intelligence all around us, and how that information is offered to all surrounding life, near and far, for those who are willing and able to listen.

Balancing Both Kinds of Intelligence: Earth Life and Technology

Like a spiral spreading outward into the world around us and beyond, we give off energy stemming from our intent. That's why our internal motivation matters greatly, as mentioned. Additionally, calmness and mindfulness have roots in our intent. In Nature, we are not viewed as superior but as members. Our diluted sense of importance only keeps us struggling.

We exist as a piece in the grand design that is all of Nature. Nature operates through multiple forms of knowing. Nature is empathic and intuitive. Nature is supportive and always seeking balance. Nature might know more about us than we know about ourselves. This is Nature's way.

Trusting our senses allows us to enliven, welcome, and embrace interconnectedness with life throughout Nature and one another. Given that life throughout Nature lives and knows the world through extraordinary abilities, we are known, too.

The physical locations of our everyday worlds—around our homes, yards, neighborhoods, and communities—matter most. This is where care may be missing; we can start here. In our areas of *home*, knowledge is indispensable. We do not need to leave behind modern life, but we do need to evolve in balance with reality, and that involves our inseparable relationship with our brothers and sisters who are Nature, collectively. Adapting our lives and balancing technology and modern "conveniences" with natural, Earth-focused life requires thoughtfulness. We must include Nature's needs and story when we consider anything before action is taken. Without our inclusion of interconnectedness, our health and world will continue to decline.

Knowledge and wisdom are prevalent throughout Nature because of individuals in Nature. The more aware we become, the more apparent it is that much intelligence exists beyond our abilities. But we are needed, too, in our own gifted ways. We may believe we are limited, whereas life in Nature is limitless. Any limitations we perceive exist only by choice.

Dung Beetles lay their eggs in manure and then encapsulate them, creating a small ball that they roll from east to west until the young hatch. Female Beetles use the sky as direction. They specifically need illumination from the Milky Way on clear, moonless nights. What looks like a dance atop the created ball appears to be a way of mapping the sky—creating an image within to find her way, and then rolling with it. Brilliance.

Life Is Pushing Back

As life throughout the natural world *knows* trust and respect, within this powerful boundary duo, boundaries reign supreme, and communication is substantial and varied—along with knowledge and vital information—all of which is known, conveyed, and experienced through a vast number of senses and emotions. This is where we intersect in interconnectedness.

These systems and individuals work together to create stability. When we, as people, cut off our true and numerous senses, we limit our interactions and information, numbing our emotions and impeding our health and happiness. We have broken the rules with life throughout Nature and with interconnectedness. Stability returns for each one of us when we begin with what is in our hearts—considerate motive and intent.

Only we can answer for ourselves, individually. Perhaps that is part of our personal struggles. We tend to refuse facts and lived experience because another looks and acts differently than we do, even when our needs are the same—because we do not want to adjust what we take to be ours in order to include what another needs. We believe we are "better." We believe we do not need these lives except for our enjoyment. Yet, because of species-specific, highly developed senses, our world would thrive if we collectively understood, appreciated, and embraced our shared strengths. Our inability to accept the worth of others impairs us, our evolution, and our future on this planet.

Life seems to be growing angry, reflecting an emotion we have projected onto it. People push boundaries in ways that are not aligned with the natural world. We have pushed hard, long, and far beyond what is sustainable, and now life is pushing back.

Go within to change our world. Become part of this brilliant life force within our homes, communities, yards, and neighborhoods. This is where our attention is needed.

Nature shows us, time and time again, *that it is not all about us.* We need to realize this soon. What an awful sadness it would be to live without hope for connection—desperately wanting to see others, knowing they are so close, yet unable to be with them—while Nature actively waits for us to connect.

Lemon and Crows

Animals in our backyard grow to know one another. They are all there together—eating, drinking, some sleeping, some learning, others playing. Animals and Birds play together.

Young Crows have interacted with Lemon, one small Rabbit. Lemon and Lemon's sibling were raised in our backyard. When I first met them, Lemon's sibling was almost twice as large as Lemon. As you can imagine, Lemon was the size of, well, the fruit.

Lemon enjoyed spending time in the yard with us and would often eat beside us. She became familiar because she grew curious when she was very tiny, watching me fill feeders. She was always watching, sensing closely. We let her be. Lemon would often rest in sunshine.

Lemon and two young Crows would interact with sticks, rocks, and other Animals. One day, each young Crow was taught about sticks by an adult. They learned to pass them to one another. They would stir the birdbath water with them. They learned together and would play. In moments, they would go after sticks together in exchanges I can only understand in human terms—a pulling back and forth of the same stick. They learned about fairness.

These Crows would also play with baby Squirrels left in our yard for short periods of time, jumping and chasing one another in bursts. I watched these Crows and Lemon jump and hop around together. They were playing. Lemon would often lie in the dirt by the house and sleep while these Crows walked around or rested.

These moments leave me only wanting more time to simply be in their presence.

Conversation with Crow

Several Crows were around one morning before noon. They have visited a few times. Sometimes they call while I am inside, and when I toss out a few shelled walnuts, they watch and then fly away. Sometimes they gather them. This time, when I opened the door, one flew to a nearby branch. I tossed out a couple of walnuts, and two other Crows came to the same low branches.

I stayed outside and slowly walked to our seed cans, preparing to fill the feeders. As I moved slowly and calmly, one Crow flew up into Tree while the others took the walnuts. She or he—I will call this individual BlackOne—then flew to another Tree directly above the yard and watched me. Then BlackOne glided down to our fence, landing opposite me, turning to face me. I was back near the cans. BlackOne was about twenty feet away on the five-foot-high fence. I looked around, and the other Crows were out of sight. Understanding a little of their ways, they remained nearby, but I could not see them. BlackOne was now facing me close to eye-level.

BlackOne began making soft cooing sounds before landing on the fence. This was new. I assumed the sounds were directed toward the other Crow family members, but BlackOne continued making them while facing me. There were also soft clicking sounds, slight puffing, and downward glances paired with the gentle cooing. Then three caws were added into the sequence. All I know is that there were no other family members nearby that I could see for these sounds to be directed toward. BlackOne was facing me, watching me, following me, and I am almost certain this Crow was trying to communicate with me in a caring, affectionate way: Family. Love. Something profoundly important.

I did not stare. I bowed my head often and placed my hand over my heart because it creates calmness within me. I bow my head as I slowly blink and open my eyes while looking downward so I do not stare or appear threatening. It is casual and mostly unnoticeable to another person. I feel this softens focus, which can make some uncomfortable, while also conveying that I am relaxed and calm—because I am. I speak no human words. Maybe BlackOne was replying. Regardless, it was an acknowledgment of kindness and information being shared about what was being felt.

Later, I did a little research and learned that cooing sounds accompanied by these movements are believed to be signs of affection. So, it was affection—possibly, for what I can know myself—and perhaps used for individuals beyond a mate. It felt as if BlackOne was replying. I understand senses more deeply; I sensed gratitude. I am so grateful. Our conversation moved me deeply.

Years later, in another state, I met another individual Crow. During our second meeting, Crow flew to one Tree above my head while I was walking and began communicating in this same way. They were looking at me with no other individual nearby. They recognized me from the week or two before and greeted me again.

These moments touch me in such a special way. They are deeply personal.

The capacity for pure, kind connection and community with individuals outside my door and within their natural world is moving. This wholeness feels right. These moments—and my time with BlackOne—will forever remain close within me. I become deeply pulled back into these memories. These experiences will never be lost on me. *Gratitude.*

Nineteen
Empathy Is of the Courageous

Empathy Is a Form of Intelligence

We could learn a great deal about empathy from life in Nature. Wisdom within empathy—flowing around, through, and together with all life—is a form of intelligence.

Few people are inclined to be highly empathic. And even those of us who have learned to rely deeply on empathic intelligence must also remember that empathy is not ours alone. Empathy is part of our human existence on Earth with Nature.

We have grown lazy within our modern worlds, our minds growing just as lazy. Given our current struggles, difficulty with acceptance, and fear, anything remotely communicative can be perceived as confrontational. At times, we are left with only a strong desire to speak and be heard without question. So we turn on one another for stimulation and belonging, when we could instead be learning together how to improve our lives and worlds with each other. Perhaps it is as simple as enjoying one another's company.

Many of us swim effortlessly in assumptions, weakening our ability to recognize our own boundaries, let alone begin to honor those of others. Too often, so many of us do not seek clarity, overlook the positive, thrive on the negative, and project it on those around us.

But when we work together, we *feel* for one another. We also feel Nature's pain, their joy, their desire to be happy, and their constant support and balance—because we are connecting with them through the senses. We *already* feel what they feel, but some of us are more sensitive to the subtleties than others.

Look to Nature not to exploit—this will ricochet back to us, and this is happening already—*but learn together with* these lives, interconnectedly.

Solving many of our modern-life challenges includes placing more focus on the unity of all life with Nature, where inspiration is supported and blossoms. It is not only *how* these lives throughout the natural world work together, but also *what* they know. Their abilities, unique intelligence, and senses are profound. *With* them, we may begin to know what it might be like to find common ground as a world community—one that includes people while honoring personal boundaries, the health of the whole, and the inclusion of these lives within our decision-making processes, without burdening this understanding with human-induced rivalry and strife.

Feeling empathy links us deeply within caring relationships. These connections matter to our health and happiness. They connect to our brain functions through the sense-emotion-thought relationship. Our senses flow through our interconnectedness with life in Nature. When we honor this kindly, Nature responds accordingly.

Looking to Nature as our source of strength and support is nothing new. What we witness is valid and real. Empathy exists. It is not unheard of or ridiculous. Honoring these relationships and their worth in their own right would do a great deal for every one of us. Given our physical make-up, the structure of our bodies and brains, and how everything connects through the senses, empathy opens us up to one another, interconnectedly. For us to better understand empathy and our place within interconnectedness, we must look to and learn from Nature as considerate, mindful guests and observers. To do this, remove assumptions and *feel what is being felt*. Begin to see and feel what it is like for another. *Not what we think*, not what we have been told, nothing built on assumption, but *what we feel*. In that space, the other might then turn the mirror on us so that we may begin to see and feel for ourselves.

The mirror of Nature reflects everything. Trust reigns supreme. When we develop our trustworthiness with the lives throughout Nature—enter: empathy.

Kindness Is of the Fierce

Perhaps we have some fundamental assumptions wrong—the ideas around "strong," for example. Maybe the strong hold certain intelligence that allows them to be *adaptable*.

To be adaptable, one must be able to work well with what they have and with those around them. That is what makes them strong.

True, sincere, pure kindness belongs to those who are strong within their sense of self. And kindness with life on this planet reflects our kindness we feel for ourselves. For again, who we are within radiates outward to life in Nature. *Kindness is of the fierce, the powerful, the strong.*

Kindness—the type that comes without need for recognition, reward, or condition—is a powerful energy that can only come from those who stand behind it fiercely. This strength—the strength of kindness—flows from those who are strong within, to the core. Nature is well aware. Here, we meet on common ground.

It is *the balance between all individuals* that creates the collective whole. Why would we seek to dominate another, or believe that appearing "stronger" than someone else is the same as being strong? To be *adaptable* is to respond to others and to the world in which we live. What are we missing from our lives, and what is missing from the situations we are in, that leads us to act in such harmful ways at times?

It is often easier to hold on to resentment, anger, and fear. It is an easy route to hold grudges and talk about others rather than practice kindness and speak directly with them. There may even be a perceived ease in attacking others for personal gain or satisfaction—but it is not grounded in strength; it only reeks of insecurity and fear. Cruelty welcomes further harm into ourselves, our relationships, and our mental space, weakening our ability to remain present.

Cruelty is not strength. This kind of harm creates deeper pain and more difficult outcomes, spreading outward, billowing in waves that eventually return to us.

Learning from Nature about the strength in kindness involves understanding the outcome of our actions and what we are responsible for—specifically regarding the lives throughout Nature. Our actions within these relationships might symbolize the actions we take toward one another as people.

Kindness might take time and understanding. It is nothing of naivety; kindness includes considerable intelligence and a willingness to grow.

Kindness is not for the selfish or faint of heart. We do not need excellent memorization skills to embody kindness. Do not be fooled; kindness is not easily manipulated. There is an *awareness* with kindness that sees readily through intentions.

Areas within each of us that are in need of kindness often require the strength to seek new knowledge in order to form a new understanding. Adaptation and kindness often join forces to help us evolve.

Some may believe that convincing examples of Animal and Nature teachings are rare. I have trouble understanding this. All lives close to home, and the time I have spent outside in Nature, have been teachers of kindness. Not only do they teach and learn through kindness, but the individual lives throughout Nature *remember*. Kindness always spreads in the most powerful ways I have ever seen.

We become known by our actions. Much of who we are results from our actions. Above all else, we are guests in a world and culture we know very little about. We cannot "hear" another if our time is spent with little respect.

The Confidence in Kindness

Kindness has a calming quality that is soothing physically and mentally. Something realigns in our moments of kindness with other people. It can be felt and sensed. Genuine kindness carries a powerful strength as well—one that is present in the quiet delivery of peaceful moments and gentle gestures. Life in Nature takes notice and feels this.

With Nature, the ripples of interconnectedness—highly functioning and very much alive, even if lost on many of us—seem to spread the message of kindness quietly and peacefully. As these moments turn into days, weeks, months, and so on, those who turn their backs on kindness seem to be in a different world, far from life in Nature. With Nature, we are known for our strength: unassuming and without need for attention. We turn to Nature, and a representative steps forward.

Crow and Squirrel have acknowledged my kindness recently through gentle moments of being together—passing stones in generosity, making soft sounds freely in our patio space, and being witnessed by Trees.

Crow came to our small patio area behind our home and called from the roof as I was working with my husband, adjusting soil in pots. Crow called from our rooftop in low, guttural, throaty sounds while looking down. Before I could respond, they flew away. My husband was astonished; he had never heard this vocalization. I have—from this individual at this location two times before. It was so unfamiliar at first that it took me a while one day to locate where the sound was coming from and who was creating it, until I felt a quiet inner prompting: "Look up."

My husband told me that when I went inside, Crow made softer sounds before leaving.

I began to understand. I had placed one nut on the front porch railing so Crow could access it if they wished, since the backyard space was very enclosed and uneasy for them. For months now, I have seen them fly past my office window while I work from home. I have heard them across the street on the opposite roof, calling. When I look, they are not there. Just a few days ago, that one nut was taken by someone—Crow.

Not too long ago, I was preparing green tea and began walking up the stairs to my office. Without thought, I felt compelled to look through the small window by the front door. I stepped down and headed that way. To my surprise, Crow was there, looking toward the window. Crow flew up to the roof next door, so I placed a few nut pieces and let them be. They did return. This was what they wanted when communicating from the backyard—a simple way to receive a little healthy food in a place that felt safe, if I was willing to offer it. A respectful and trusting relationship built from kindness.

None of this is unusual, really. Life is aware in ways that are less apparent to us, so it would seem. For them, it moves softly through the spring breeze, touching all. I do believe I am known now for the kindness I place front and center with the world beyond my doorstep, in trust and great respect, over the last fourteen years.

I have never claimed to be perfect, and I never will be. I am not interested in perfection. I no longer wish to work so hard at human relationships, either. I have focused great attention on becoming a better individual where I stand on this Earth, in my daily interactions with those around me—especially life in the natural world. That day with Crow, there was a shift within me.

My relationships with the lives I know and have known in Nature have allowed me to trust myself, and to form new relationships with people and my work in a different way now. I can feel it. The difference feels lighter. It feels more integrated—easier. A couple of individuals very important to me have communicated honestly in meaningful ways where, in the past, things were very different with others from whom trust was expected. I have walked through a door and closed it, revealing so much possibility among sunshine and fragrant beauty.

My relationships with the natural world have now come full circle and include some of the most wonderful people, too. Kindness—the kindness I have cultivated from within my inner Nature and extended to life in Nature—is now touching and engulfing other relationships as well. Life is healthy and strong. Kindness is the source of this strength.

I smile a lot these days, and I even noticed this morning that it feels more comfortable— perhaps more natural—to smile than not.

Twenty

We Decide

Pause Modern Conveniences

I, alone. I write this to remind all of us that we are never alone within ourselves, interconnected to all life on this planet, both human and nonhuman. It is our decision to accept this or not accept it. And it is ours alone.

Our attitude, ability to remain open, and willingness to set aside assumptions of what we know, what we have learned, have been told, or have been taught, allow us to see what it is that Nature has to share with us. Without pressure or dominance, but with gentle support extending from our respected, calm distance—an honored presence—we gain a glimpse into a world we assume we know when we really know so little about. Truly, it does start with us. Nature is waiting, quietly.

Pause on modern conveniences and distractions that separate this part of our world from life in Nature. There is a time and place for these things, too, but not with the lives who appreciate the intricate balance of trust and respect. Be mindful of this, and mindful that anything beyond our home is not our home. It becomes ours when those who call it home are acknowledged as the keepers, and humanity allows for that space.

Taking but a half hour a day to sit quietly with Nature, observing without words or thoughts of our own lives, will bring a connection of support and comfort. Independent of computerized devices, we decide our mental and physical states because they, too, are welcomed in the world of Nature. We decide to stop trying to prove ourselves and instead simply be.

No one can make us know or believe in an experience unless we experience it on our own and hold it close, for that is what it is meant to be for us. But all of this depends on us. Kindness and respect must be front and center, radiating from our Nature within.

We can become so caught up in our own needs and desires that many signs and signals for connection from individuals in Nature go unnoticed and even ignored. We need to make the first step. We need to decide to stop and listen.

Listening happens with our ears, yes. But for the communication that happens in Nature, hearing is but part of such a larger communicative network. Communication happens through the vast number of sensory channels that connect these individual lives—Plant, Animal, Insect. These signals are carried great distances across ocean currents and streams, through the cool mountain air that meets the heat of the Front Range blowing across the Great Plains.

We seem to need instruments and specialized scientists to interpret information around us. Nature, on the other hand, has evolved to rely on this information and use it—the light of the sun, the temperature of the day, perhaps energy in all its varying forms—as a guide full of intelligent knowledge that intelligent minds know and understand. Superior to our abilities, definitely, and we can decide to approach these lives with appreciation for this because it is nothing short of brilliant. On common ground, we may receive a glimpse, "hearing" through other ways of knowing what life in the natural world hears, too.

In Nature's search for balance, we can begin our own balance, too. But we need to decide to have this in our lives first.

Nature experiences play, joy, love, and fear just as we do. Even among us as people, our expressions vary according to the individual. What differs is the expression of these emotions through our unified senses, varying from individual to individual. Again, this holds true for the lives throughout Nature. The same holds true for many experiences and senses: We can decide to "listen" and understand the balancing act carried throughout Nature on this Earth. In deciding to embrace Nature's intelligence, we find support, experience play, know unconditional love, and find our purpose. Perhaps it is we who have something to learn.

Regardless of our experiences and our history, when we dull our emotions because we wish to avoid them or have negative associations with them, we are also missing an entire way of communicating, learning, understanding, growing, and gaining intellectual insight and experiences. We are removing ourselves from this greater, unified source of knowing and intelligence. We have removed ourselves from all that is Nature.

We can decide for ourselves, based on our own direct experiences with life in Nature, while respecting their boundaries, needs, and goals, so that we may begin to experience, witness, feel, see, and learn from these lives and their brilliant intelligence. In return, we open the door of willingness so that it may remain open.

Interconnectedness is within life in Nature. Through our experiences, the senses are triggered or activated, and how this translates to our physical and emotional selves and our minds together, we begin to feel. We see a glimpse of what it means to be one.

Remain open to areas of learning and understanding ourselves and others. Doing so with life in Nature builds mutual trust. It is here that we can find relentless strength and direction for all the other areas of our lives. There is so much more going on all around us beyond words.

We decide our limitations, and we decide our health, happiness, and acceptance of our need for interconnectedness. If we need support in learning to let go of our past decisions, life in Nature will be there for those who are worthy of trust. Life in Nature will aid in removing our ties to inaccurate understandings about them and of them. They will embrace ourselves as we are and for who we are if we can mirror them in their far superior abilities to maintain healthy relationships. Even among those who need one another for food, trust and respect live. Nature is waiting, quietly.

Into My Eyes, Quietly Understood—Seal, Squirrel, Chickadee

How much does our current connection to life in Nature influence the answer to the question, "Who am I?"

Recently, we were kayaking on a bay. Harbor Seals are known to visit for brief moments. This Seal stayed with us on this day, swimming below us in slow turns of the body, back and forth. They stayed with us as they popped out of the water, head exposed, to breathe. I could hear the opening of the nose for air, along with the exhale and deep inhale—water spraying. There was such detail of so many whiskers shining in sunlight, and I could see deep into his or her eyes. Only about three feet away at times, this Seal stayed with us for about half an hour. My deepest connections and learning usually happen when I am not with other people. The talking that happens, naturally so, detracts from my experience to "hear." Knowing myself well, I know to understand more in later, quieter moments; moments of quiet return where I can feel it out.

The following week, my husband and I were walking in a community park. It was a beautiful place close to home with a few roads and unpaved trails. There were so many Birds and Trees—tall, old elders of our forests. The huge Ferns and smells are of times that seem long lost. If I can imagine, traffic not too far away can be confused with the approach of a Dinosaur walking through these mighty Trees on this same Earth.

As we headed back to the car, there she was, waiting. We were a good distance away, and I could see Squirrel on the post directly next to my car. She turned toward us as we approached. This is not a park where people would leave food behind or drop scraps. Food sources are abundant, and these lives seem to thrive.

Even after all these years, my brain went to the very simplistic thought that this Squirrel wanted food from me. I very rarely have food on me when I am on a hike—never in this place we visit randomly. Yet, she watched me closely. Squirrel was looking at me—looking into my eyes, in that way where it is felt. She started to leave, then turned back to look a little longer before climbing back into the brush slowly. It was odd, so I thought. But was it? Even after all these years, was it really that odd?

Early the next week, I was on my porch checking feeders. One of the Chickadees landed on the feeder right in front of me, stopped, turned, leaned toward me, and stared into my eyes. I paused. I waited and slowly looked away. Chickadee left after a few moments.

Interactions are common for me because I am aware of these moments throughout my days. Perhaps there are profound messages conveyed that I am unable to interpret completely, given my human limitations; most of which are personal and for me alone.

What I do know is that I have decided to honor these moments by honoring these individuals in a quiet, respectful way, allowing them the space to be, even if they decide to know me in their ways that are lost on me. I would want nothing more than to be included. Perhaps that is the point. I am included and part of their world, too.

For me, what is most important is that I remain aware of each individual and their relationship with me in that moment when they seek to connect with me. These moments have happened across many years now, even up to the final moments of writing these words in my edits with my publisher. They have happened because I have decided that each individual life matters on their own terms, and I will respect them because I have grown—evolved—into this oneness with them. In these moments, each individual and I become one in trust and respect.

Part Four

We Evolve to Respect

Twenty-One

Being and the Individual

The Uniqueness of the Individual

As people, we speak of our uniqueness and how exquisite it is. Yet these expressed words are often not honored—many are tormented and bullied for their uniqueness. Our right to be—to exist and contribute to this life and our world—radiates from our inherent worth, our innate sense of self, our natural being. We are each unique individuals of strength that our world needs, together. These are the ways of Nature.

Nature is full of individuals, with their own families, sense of self, and communities trying to live well, just like each one of us.

We are all someone. Life throughout Nature is gentle and kind. We, too, are known as the individual we are when near their company.

With these individuals, we can learn and embody our strength when we offer kindness in return. Nature teaches us to look in the mirror—to embrace the worth of individual life in the natural world as they are, while also embracing our own reflection.

We cannot forget that Nature is beyond trails and beautiful vistas. Nature is life around our homes and neighborhoods. Nature is each one of us, too.

These are not feel-good words. *This is life on Earth*, and we are fortunate to have such diversity, such uniqueness.

Nature is all of the unique, individual life in our homes, our Pets, the Animals we care for, and who provide so much for us.

The depth of character and emotion in life is fluent everywhere. Each life is different, and we see those differences when we allow ourselves to *see* and *know* clearly.

If there is a way to cause struggle, there is a way to embrace and create balance. Master the art of learning from life and working with the natural world, because strength is found in those who seek alternatives that are for the good of the whole—and we begin to feel better in the process. Lights of inspiration begin to flicker in our moments together, and emotional connection is automatic and necessary. Welcoming these lives touches us immediately.

When we accept ourselves in our state of being—who we are and what we are in kindness and hope—we see others clearly and appreciate uniqueness.

The differences among each one of us do not stop with us as people. Quietly, Nature guides us.

Accepting the Exceptional

We are not alone or separate from life outside our window; we are not alone in our uniqueness. The same individuality, from bold and curious to shy and reserved, exists throughout all nonhuman life, from Plants to Insects. Life in Nature—all life—consists of exceptional individuals with families and support systems. Each unique individual matters in their relationships.

When we intervene in ways that disrupt their families, support systems, and understanding of a functioning community, it negatively impacts our communities. Remaining mindful of this creates an acceptance of their exceptional contribution, and we become open to knowing the same of ourselves.

In our embrace of individual uniqueness, we instantly remember that individuals in Nature are their own beings and form their own cultures and families. This alone is worthy of trust in how they express and experience events and situations. In each of our worlds, we matter, and in each of our worlds, we exist together as part of the greater whole of life.

Hippo and Ostrich

Beyond what we use Nature for to exist, even when we do not acknowledge the individual in honor of their life and gifts, generosity is the exceptional way of Nature.

Hippos have offered some of their own food to neighbors held in captivity in zoos. I have seen them gathering feed grain in their mouths, opening their strong, crushing jaws to a neighboring Ostrich who waited patiently. Ostrich then began to peck the seeds from their considerate and compassionate friend over the small wall that separated them.

I have witnessed other lives, knowing only their neighbors, who have also done this time and again.

As people, we understand this—it is touching. We might not see a smile that looks familiar to us, represented by a mouth turning up at the corners, or hear a verbal human "thank you." But because we do not witness what is familiar to us does not mean these two individuals are not expressing a great deal between one another; perhaps love and friendship. Maybe they bonded over histories and stories expressed to one another in ways we simply do not understand. Relationships in Nature are as unique as our own and exceptional in their depth and personal value to each of these individuals.

Honor Life in Nature

If we believe we are not brilliant—if somewhere along the way we were made to believe we are not exceptionally brilliant as we are—then we cannot recognize life in Nature as brilliant in their own unique ways. The lives that are Nature are just like us in so many exceptional ways. And each individual is an exceptional being, just like each one of us.

Accepting the exceptional within each one of us allows us to come to a place where we can see, know, and embrace it in others. Nature supports us when supported, and the slow cracks that have been widening begin to lose their divide.

Steps made to honor life in Nature never go unheard, and certainly not within our bodies. Be okay with a different approach that is positive and supportive. Be willing to change if we want a better life and world. Because what matters here, for individual health and happiness, is that it is not only about humanity. It is about us as a whole.

Insects and Their Families

Insects are given so little of the benefit they are due. I am always amazed at the abilities of Insects to find what they need so readily, while we struggle not to get lost, even when using GPS at times.

In arid climates, during the endless heat and drier-than-dry months, a little water in our yard—sometimes a very small puddle—is found by Paper Wasps and Mud Wasps wanting nothing more than to build their nests to help the whole of their communities, which is their part in Earth's larger community.

We would be in deep trouble without these lives of Insects. I do not think we realize how important they are beyond their role in our consumption. We do not realize why it is so important not to pollute and poison their worlds—the worlds of other lives in Nature indirectly—and how it all comes back to us. What we think of them holds some reflection of ourselves. We are poisoning ourselves, too. We, as an Earth, deserve better. We need better.

Embrace Balance

Embrace balance and boundaries while also offering decency and acceptance. In turn, begin to see these same needs within our worlds, too. And the support offered from these lives greatly resonates around us. Just as our experiences have impacted how we cope and handle situations, the same holds true for life in Nature. The lives working together in the natural world work with change and adjust to circumstances so much better than many of us do. This does not mean they do not feel deeply. Some do not cope as well, just like some of us when faced with horrific situations. They also work to overcome what they have witnessed and lived through. Plants are not the exception either, and we would not be here without these amazing individuals, the many brilliant Trees and flowering individuals.

Plants are quite different from us as humans, and from, say, Mountain Lion, Otter, and Turtle. There are good reasons why they have developed their own unique ways to survive as individuals and within communities. Plants cannot run. Much of who they are helps them thrive. All life needs Plant life in basic, varying, and wide-ranging ways. We cannot choke out what is needed. Plants are individuals full of life, understanding, communication, response... *awareness*—aware of more than we presume. I have found Plants in all forms to be incredibly supportive; intelligent in their ways and in what they need, in conjunction with the life around them. There is something for us to learn here. Remain open to understanding and knowing another. When we come to know and breathe the value in doing the same for ourselves, our bodies and minds find balance, too. We come to understand the importance of this health within all of our unique relationships. Plants, including those on your windowsill, know you.

We have created problems for all life on this planet. Perhaps each one of us, from today going forward, can start to do things differently. Sophisticated as they are, we can learn from and appreciate the worlds around our own homes by learning from unique individuals in the light of a beautiful day. In our precious spaces in our homes and surrounding communities, we see more, learn more, and understand more. We see more in ourselves, learn more about ourselves, and understand more of ourselves. We can learn about weather and situations, relationships, and struggles. We learn what is most important in healthy relationships through those with individuals who are living in health and balance.

In supporting individuals in Nature, we begin to feel supported in return—a closeness. We feel it in our minds, in our bodies, in our breath, and in our hearts. We become excited to continue the balance with them in our neighborhoods and communities. We begin to know we are all connected and, together, we are not alone—we are exceptional, together with Nature.

Knowledge from the Uniqueness

There was a time when individuals in Nature had rights of their own and were revered for their gifts, strengths, and talents as a way of supporting life on Earth. It was a guide to living together. The belief was that there was no bad in the gifts bestowed, only opportunities to learn and grow, to become better people in support of other people as part of this larger community of life.

There was uniqueness in the ways of Coyote, Deer, Badger, Ant, Porcupine, Dragonfly, and so many other species. Their ways were learned from watching and honoring these lives as individuals over time. During this time, a relationship was built, and from that relationship came understanding.

Life in Nature cannot help us until we let them. This starts with trusting in the individual so that they may again begin to trust those who are trustworthy.

To remain clear: trust is not about contact and physical closeness. Trust is what *they* need to know with us. They need us to honor their boundaries. Trust will only take root in genuine care and concern within the relationship; in the well-being of others, in the right of an individual to be as they were meant to be when created for this Earth.

Our Nature within—our cells—deserves more kindness, too. After all, we are either in this together or not. Life in Nature is not hateful in the way humans have become. If any act or way in Nature appears hateful, perhaps turn to our mirrors and ask ourselves why. On common ground, free of the dumping of our own issues and broken trust, life in Nature already aids and heals. Our Nature within responds accordingly.

We have pushed Earth too far from her center. Earth has given us every indication that she is trying to right herself and return to balance. During all of this, we are fighting for our health and fighting for happiness. We *resist* the path to an evolving self, learning and growing, and having purpose while loving our commitments and responsibilities toward the community of life around us—and we do so by our own choosing. Life in Nature is falling off the edge and, in that fall, we are being pulled along, too. We are to blame. We are to create change, too. We can progress, but in that progression, we have to pull back. We can have balance *and* progress. If we are as intelligent as we believe we are, this is within our reach. And for those of us who struggle with change, the alternative will be far more abrupt, like a worn tire suddenly erupting while we speed down the highway.

Our basic need for interconnectedness—honoring our natural world around us and within us—begins when we help life in Nature as they need. In this help, what we need is mirrored back in a healthy, deep breath. But we need to restore respect where respect once was with these unique individuals.

In Our Pause

Once we evolve in our relationships with the natural world, we begin to give these lives credit. Here, we begin to open new doors of understanding. We begin to feel better, and we start to want those around us to be better, too.

Finding meaning within our lives is not ours alone. Meaning is shared experience in some form that creates good for the larger whole. So, when we begin to see each life in Nature as an individual within their own world and right, something within us shifts. For this shift, we need to step out of ourselves to know the bigger picture of life on Earth. *Our Earth, together.*

Allowing a life in the natural world to *be*—to exist as they need—for who they are as an individual, and wanting them to be well in support and kindness, including what is good for them, opens doors within us to a greater good. In our moment of pause, in a simple act of respect, change begins.

One of Our Family

After years of interactions, in subtle moments and many experiences over time, these lives know one another across species, and certainly, the bond runs deep within families.

Some people could learn a great deal about love, trust, and respect from the natural world. It benefits all of us to know what it means to honor and value these healthy relationship attributes that are so vital in human relationships, too. Life in Nature is our support if we are mindful enough to create the space for that support.

I have witnessed Deer avoiding areas for many days where loved ones had been killed by cars. I have watched a very young baby play with their sibling the day before in our yard while three adults watched from the side. They run freely, just as Dogs do. They chase and turn in safe places to enjoy some adult-supervised, carefree moments. They offer the same displays of joy and play that I have seen in Squirrels.

That baby mattered to that family. The same for the Raccoon family, whose member was hit by a car, bleeding and suffering in the dark street—something I wish I had never witnessed. Watching another family member reach over, searching their injured loved one's face for life, wanting and trying to get them off the road... People do not realize how blinding headlights are on roads, especially those that are very dark.

I have been caught in these bright headlights myself when I stopped and jumped out of my own car to grab a little Dog who was lost and confused on the street. As I stood and turned to face my car, my own headlights blinded me. I could not tell which way was which. No sides of the road could be seen, nor what was above me. I could see nothing but bright light.

I say this not to create a heavy darkness, but to bring light to the fact that life throughout the natural world is so much like us. Grief is something we have all experienced, and it is felt more deeply than words or actions piled heavily upon it. We tend to understand grief more readily.

Just like us, Animals suffer greatly from loss. People must begin to understand this through strong, compassionate, and intelligent feeling, and offer them the emotional space they need. We must do a lot differently and, in doing so, we will become better for it. Out of consideration, out of respect, in learning this state of selflessness, we open our inner door to the empathic world at large. We can feel it. And in our compassion, life throughout Nature appreciates our acts, thoughts, and state of being more than words.

When we have moments of this interconnectedness that touch us deeply, we walk away more grounded in our own uniqueness as the individuals we are and want to be on this sacred Earth.

Twenty-Two

Honoring the Process

Calmness and Presence

Here, we focus on the present—today—and each day moving forward. Reflection helps us process and filter our experiences. Learning is vital to change. Remember, we can create our own positive and supported life and future. Where there is the "want," there is a way. Be attentive to our thoughts, opinions, and actions with confidence and kindness, and create that world around us. Regardless of yesterday or over thirty years ago, focus on the now. By being present, intentional, and trustworthy, we can cultivate reciprocal relationships with Nature that heal both Earth and us.

Regard the necessity of our own process by honoring the process of the natural world. With life in Nature, this silent declaration of our intentions and what motivates us is communicated not through human language, but through all that we emit through our bodies, senses, and considerate energy. *Calm and focused.*

How is this even possible? We focus close to home—on our street and in our neighborhood. Looking up at the sky, we go within. Remain present within our bodies, our brains, and our minds. This helps keep us focused on the now. Calming our thoughts, assumptions, expectations, and inner chatter is essential.

Meditation is a powerful and empowering way to learn to be present. It helps us remain open and release our thoughts and words so we may know through other senses. As we stand in our mindful state, we are what we are within. With Nature, we can also learn this directly from life around us.

With the individual lives in Nature close to home, calm the brain and quiet the thoughts, and pay attention. Quietly, be with what is going on through what is sensed. Begin to feel emotion. "Hear" another through other senses—through intelligence beyond sound. Be aware enough to know another without injecting personal thoughts or assumptions into the moment. Simply be calm, tranquil, and present with life in the yard or with a Houseplant on the windowsill. Feel and sense emotions through the peaceful and relaxed body. Look with your eyes, listen with your ears, but allow supreme space for the rest of the senses.

Let images flash through the mind. With Animals, avoid staring. Soft, slow blinking while looking away during the moment the eyes are closed can help. Open the eyes while looking in another direction and blink again. With a gentle gaze, return quietly to the one with whom you are interacting in the moment.

Take the time to begin to feel what is sensed with another while honoring boundaries. Perhaps simply be together in quiet with care and love held within. This is appreciated deeply, too. Listen without personal bias. These are ways to know others without a word spoken. Some of us have honed these skills. Most of us need only begin. Inhale and relax, exhale deeply—calm and focused on calm.

Intentions matter. Our purpose is to learn from, through, and with the life around us. Perhaps, in fleeting moments, this means giving them space and returning another time. Our time together is never about one-sided wants. We echo interconnectedness when we give another in the natural world what they need now. This individual in Nature will feel our intent and, when it is positive and considerate, may offer the same consideration in return.

Desire Aligns with Evolution

For positive, productive relationships, we need to want positive, productive relationships. If we need something to change, we must want to change. Even when we feel a special connection or camaraderie from the beginning, depth comes through moments shared over time.

Take the time to learn from relationships with individual life in Nature. If necessary, we will be gently guided to find ourselves, learn about ourselves, and feel supported in who we are and who we wish to become. Inspiration extends from our desire and ability to be present with life around us in the natural world.

Offer yourself the self-respect of time to be who and what you need to be. To heal ourselves, we need to heal with the natural world around us. We are designed to evolve and grow together. If we are fortunate enough to have a path before us in life, embrace it. It is a gift afforded to us—to learn, become wiser, and grow in understanding—a gift not granted to those taken too soon. Rarely is our course entirely clear all the time.

When we take others for granted—perhaps taking advantage of relationships that could be so much more loving—those actions are ultimately rejected, sometimes painfully, both within ourselves and with those around us. If something hurts when released, it will not feel good upon its return. Actions and memories stored within can fester like toxins, creating inner disruption. We can choose differently—to be happier, to honor our responsibilities, and to make wiser choices, including those tied to purpose. We need time, space, and the willingness to evolve. We need to want this.

Life in Nature does this effortlessly. Through balanced learning from one another—me, you, and the individuals throughout the natural world—our internal struggles begin to realign and, as a whole community, we begin to heal. Purpose starts taking form in the evolution we now find ourselves within. Positive contributions begin to add our own special touch to what supports the whole.

As part of the equation on planet Earth, we have lost our way. Nature continues to adapt—sometimes successfully, often not quickly enough because of human actions and inconsiderateness. We are not immune to the latter, and it is more evident now than ever. Nature has been trying to right this for a long time. Earth, as a collective force that is Nature, will find balance. That is Nature's way. And it might not be to our liking.

Committing to a devoted half hour each day to be with these lives that are Nature—in positivity and kindness, with their needs as our main consideration—is where we begin. It is that simple. Our physical, mental, spiritual, and emotional well-being depends on it. Nature's well-being depends on it. We need one another—this has always been true.

Begin today. Take time now, with excitement and enthusiastically supportive calm, to sit with life in the natural world close to or within where you live. Perhaps this individual is interested in sharing a moment with you, too. Images, sensations, emotions, and any other positive support, experience, or outcome are personal and meant for us alone. These relationships are special for each one of us. This is our first step toward evolving with life in Nature.

Listening: Receiving the Quiet

To hear is one thing; to listen involves the ability to shift our attention to another, sometimes quickly and more deeply than many are accustomed to. Listening means taking in what another might be expressing, regardless of whether sound is part of their process. It involves much more than hearing alone.

Sound may not even be part of the moment. We can "listen" to the senses triggered by another. We can receive information offered in a flash of knowing, even when it is not fully understood, yet always comes from outside ourselves and reaches inward. We can listen through feeling with another, without a word spoken.

Any sound relationship needs a healthy foundation. The same is true with life in Nature. We need to take the first step toward finding common ground in our understanding of the life around us, and this includes an emotional and personal understanding. No one can do this for us, and these moments together cannot be artificial. Each one is ours alone—unique and special, mirrored back to us through the life shared in the moment.

In an unaltered, present mental state, we are truly present. In moments when we wish to learn from Nature, we owe these lives a respectful approach.

Anything other than respect prevents trust from forming between us. Nature can sense when we are present and when we are not. By being fully here, we show that we are, in fact, listening and willing to listen. As we begin to embrace life emotionally among all life in Nature, we learn to evolve with them and build a process that is our own. We learn how through them.

Begin to learn to hear Nature and these individuals through your own personal experiences. These moments are meant for you. Some may be deeply personal and intended only for you. Others you may wish to share with trustworthy people. Moments of support, clarity, kindness, and understanding reach us through our senses—not from our past, but directly with life in Nature.

These relationships must be entered through the heart and with kindness. This is how we receive the quiet. Be still. Respectful. Calm and present, with love and kindness held within, because these qualities are known around us even when we wish they were not.

Receive the quiet and the immense volume of communication flowing through this way of understanding. Be at peace with the emotions that arise. We are meant to balance and heal among the natural world. Peace comes. Knowledge is power, and power takes its kindest form here.

Sometimes, in these moments, we learn what we truly need—and often it is not what we thought we needed before receiving the quiet with Nature.

Walking with Nature

Natural systems allow a reconnection to occur that often feels effortless, though never lacking in reality. They open our eyes to who Nature is collectively and individually. Interconnectedness leads us to see who we are, how we operate, and ultimately what we need—or have always needed. This happens in moments that build over years: our personal evolution.

Our connections and relationships with these lives matter. They leave us feeling a sense of trust within ourselves. If we extend trust because we are, in fact, trustworthy, we are viewed positively—differently. In this mutual state of support, we begin to desire protection for these lives. Here, in our proper place within the equation of interconnectedness, perhaps our own desires evolve in reflection.

We feel less afraid and more complete. Our sense of community expands exponentially, intelligently, purposefully, and spiritually. Life in and through the natural world seeks our greatest good within so that we may become one with the inherent worth of life on this uniquely beautiful Earth. It is our choice.

The lives outside our doors wait for us to take the first steps on this path together.

Approach life with Nature. Move forward in life alongside our natural world—together. The depth of support carries us forward with purpose and conviction. In positivity and focus on our future, as one community of life on Earth, we can live in a state of oneness in our everyday lives among Trees and through the breeze.

This is the reality of interconnectedness within the equation of natural life in our modern world. Once trust is established because we have become trustworthy, respect evolves.

Twenty-Three
Nature Awakens Inspiration

The Landscape of Home

It is in the security of a space—one that is safe for us, familiar and comfortable, where we feel unjudged and free of expectations—that we find our place of home.

A landscape of home, no matter the size or location, becomes a place where we can find solace among the many lives that thrive in acceptance of self and seek balance, harmony, and peace. Inspiration stirs all around us softly, yet with boundless conviction.

When we are home, in our yards and neighborhoods, these places become more familiar and more accepting. We learn of the individual lives of the natural world who frequent the area, pass through, or call the community home, too. Here, our sense of home swells. It begins to broaden and multiply. Life around us interacts more and notices us in our acknowledgment of them and their place to be. The connections build and grow with revered boundaries, comfort, and space. Our home becomes truly our home—a place we love to be. This home fills us from our core. We are interconnected. We begin to feel what it is like to exist together.

For balance, we need to understand how to offer it just as much as we need to receive it. With life in Nature, the physical functions within our bodies, brains, and minds begin to ease and turn toward inspiration. Once in this safe place, we feel. This is a very good thing, and we feel it within our bodies, too. Our feelings—and the ability to feel for another, embracing empathy, understanding, camaraderie, and intuitive knowing—are the real substance of relating to life around us. We learn to take the time, and often a moment or two, to listen to ourselves in unity through the senses that connect us all with life in Nature and with one another.

Interconnectedness Is Present

When we learn to pull ourselves out of this shared sensory experience and observe more clearly, and learn to decipher what belongs to us and what belongs to another, this expands our intelligence even deeper. The comfort found in space with life in Nature teaches us to exist considerately together on the same level. This is where mutual understanding breathes.

Make no mistake—it is often we who need to step up, remain open, be strong, and become intelligent enough to feel with them. Yet in this strength, we learn more about ourselves and what we need in order to return to, or begin, our own balance.

Take time in the morning to ground and center yourself. Perhaps connect with your beliefs, whatever positive role they hold in your life, and feel support from the natural world. Meditation helps quiet the brain and keeps us present within our bodies—a place where we can simply be.

This is also a place for gratitude. Quietly within, acknowledge all that is good. Respond to the positive with genuine gratitude. Remember what matters most above all else, such as our health. Within our health is our interconnectedness with all life on this special planet, unlike any other we have known. Remain aware, yet quietly within.

Moments throughout our days—walking to and from our cars, class, work, or the store, or simply pausing at the end of the day—help ground us in gratitude and interconnectedness. Crack the window open. Hear Birds. See the clouds, sky, and Trees. Smell the air.

Pause at the end of the day to sit quietly with Squirrels among Trees without projecting expectations onto them. Remain aware that perhaps another life wants solitude, and receive permission first through feelings, sensations, and intuitive knowing before entering their space. Hear Owl and feel the air shift and cool. Let life be with us. Sense what these individuals may need within that moment. Hear this inner "voice" and honor it. Offer gratitude through thought and feeling, with both heart and mind. Do this actively. Do this quietly, within.

The Reciprocity of Gratitude

As people, we often wait for others to take care of situations that would benefit all of us. We are quick to want instant and easy solutions to the struggles within our lives, our relationships, our health, and the natural world around us.

But can we adjust our inner light to focus on the positive within our world and within our lives?

Feeling and becoming grateful ignites inspiration. Saying the words and truly feeling the meaning behind them are very different experiences. Emotion creates something very different within us—it breathes life into inspiration.

Gratitude becomes cyclical when incorporated with meditation and when felt alongside individual life throughout the natural world.

Gratitude permeates our cells and creates space for positivity around us. This includes our beliefs when they are rooted in kindness and community, including life throughout Nature. Inspiration reverberates from gratitude like a ring of energy reaching outward in support of others.

What is wonderful is that we can remain within our own worlds—our homes, properties, neighborhoods, and communities—and focus there, because these are the places that matter most to our healthy and happy lives. This is where we meet Nature on common ground. And this is where we can begin to heal what most needs healing.

We feel grateful. And we can return to this space as often as we wish.

The Emerging Sense

These places are ours, and since we have claimed them as home, we also have the power and inspiration to connect with the life that lives near and dear to us. When inspiration is ignited through connections we never knew were possible, an understanding of ourselves settles within us. We begin to know another in ways we never thought possible, considered, or realized were so important.

Accepting life around our homes, neighborhoods, and communities in a new, special, and brighter light not only inspires us and supports our health but also ignites our need for interconnectedness.

As we create our own safe place of comfort and rest with these lives around us, it becomes imperative that our words match our kindness. How we refer to others shapes how we treat them, regardless of whether they are human or other than human. Change begins with the energy we hold within and offer outward to others. We are limited by our words, so using limiting language only keeps us trapped in places where we feel unhappy and disconnected. Replacing "it" with words of dignity and accuracy begins to change everything.

For us to build interconnectedness in health and balance, we must remain mindful of the words we use, the energy they create within us, the feelings they evoke, the kindness of our intent, and the motives behind our actions. We create our reality, beginning now.

Now, we become mindful of how we treat our yards, the lives within those spaces, and the visitors who arrive as part of the larger community that many individuals need in order to thrive. From our yards and outward through our neighborhoods and communities, positive energy ripples outward in unity. Our basic need for interconnectedness begins to be fulfilled.

It is these relationships that provide unconditional and ever-present support, even during moments when we struggle with disconnection from other people. Our happiness, and our inspiration for happiness, is shaped through how we come to know ourselves in relationship with life in Nature. By beginning within—and this includes life in Nature, because we cannot experience interconnectedness without these brilliant lives—we begin to see humanity in a new light.

Inspiration Is Echoed, Considerately

There is something deeply powerful and empowering about being considerate. The most powerful reaction to consideration is the feeling it creates within us. The positive energy radiating outward does not fade but instead builds with each moment of similar intention. Echoing back to us is inspiration.

We are many. Humanity has outgrown the planet and strained ecosystems everywhere. But each one of us is one of the many, and if each individual begins honoring the basic need for interconnectedness, the strength of the natural world and the planet will respond accordingly.

When we are considerate of other people, they may express gratitude or return the kindness, and this feels good. Yet the most powerful thing is when kindness is paid forward, passed along, or continued in another interaction.

With life in Nature, consideration is felt within our bodies and cells in ways difficult to describe. The sensation is one of interconnectedness, continuing with life in Nature and tying us back to the source of acknowledgment itself. We become known by our actions and our treatment of others. We become known by our regard for boundaries and our acceptance of another's place on the land, within the natural community, and alongside us. All of this echoes back within and through us. We become known as this consideration.

As kindness and consideration are strengths, and require awareness of another's presence, including their state of being, even simply offering space—something we often take for granted—is a meaningful step that all of us can take to create supportive ripples.

Perhaps if we look to these lives to better understand why we are the way we are, we may very well become better for it...along with our health, our relationships, our sense of place in the world, and our inspiration.

Through consideration, we learn more about our own emotions. Poets, artists, musicians, writers, and dancers have long found inspiration in life throughout Nature. There is a reason why. We do not need to be artists to understand inspiration. We can feel it echoed back through admiration and appreciation.

Our felt inspiration is not ours to keep. Instead, we return it outward into the world, actively and with positive anticipation for more, considerately.

And if we view those with such gifts negatively, then perhaps we are the ones most in need of healing. Inspiration is awakened through consideration and is capable of creating astonishing transformation—whatever positive inspiration we may sense.

Be wise in your consideration of life in Nature. They know what they are doing. Have faith in their brilliant intelligence, emotional agility, and awareness. Leave undisturbed those who need great space around them. Provide natural areas with the room they need for the lives within them to thrive.

Be wise so that, in the process, we may discover our own worth as well. Be responsible, considerately—not for what we want, but for what our community of life throughout this magnificent Earth truly needs.

With Unconditional Love—We See Ourselves

My home has often included the streets I walked, a balcony with potted Plants, a small patio beside a small yard, a little birdbath, and a feeder.

I know now what unconditional love feels like.

Something within me has shifted. My deep love for Nature, Animals, and Insects remains with me just as it did when I was a child, though now it is understood differently. I have learned about trust, respect, and community from these lives. I now know kindness, reciprocity, and gratitude. I remain humbled by their intelligence, having learned only a sliver of their light, yet no less amazed.

I have been supported in ways I never imagined. I have learned through gentle, deeply personal moments rooted in respectful love—moments meant for me to learn with them, together.

But first, I had to become well enough within myself. That work took many years. Life in the natural world stayed by me all along the way.

There is no void in my life now. This gives me a profound sense of strength, though perhaps it is not strength at all—it is love. I am loved greatly, and I love greatly in return.

As time has passed and I have remained open with life in Nature, I have continued to grow and evolve through our interconnectedness. I need these relationships because they are part of who I am.

All of us can learn and evolve. We only need to be inspired toward the light that is ours to follow. Support for all that we are, and all we wish to become, can be found in positive, trusting, respectful, and supportive relationships with individual life throughout Nature.

Here, above the fog and noise of daily needs, responsibilities, functions, and distractions, we see ourselves most clearly.

We become more complete, better within our relationships, and inspired by what life still holds for us.

And we begin to want the best for others, too.

Twenty-Four

Embracing Inherent Worth and Accepting Empowerment

Integrating Inherent Worth

Because our Earth is a complete community connecting different populations, each individual holds inherent worth within their own state of being, without any value assigned by the needs of others. In other words: together, we all carry our own irreplaceable worth.

When we apply inherent worth to our own lives and to those we care for and love, we begin to understand deep intrinsic importance. When we apply inherent worth to life in Nature around us in responsible, balanced, and trusting ways, we embrace interconnectedness within ourselves. Encouragement then flows back to us from these lives, growing as we continue and spreading through Nature in profound ways. We become known as this among them. Perhaps we become one with them.

Once inherent worth is integrated into our awareness, we are empowered through forms of support so deep and rooted that we may have never experienced them before. We begin to emit inspiration and balanced wellness felt by all life.

Do not explain this away, dismiss it as childish, or ignore the magnitude of what it means for each one of us. Power and wisdom are not earned through force or taking. They are realized through honoring the interconnectedness of all life—through understanding that each of us is unique in our own way and that our existence is worthy because we exist, and that we are each one piece of a much larger whole.

We can do extraordinary things when we consider our world from this wider perspective. Our family expands greatly, yet without burden, and with unparalleled support, healing, inspiration, and intelligence. This becomes our wisdom and Earth's support for the empowerment of our uniquely individual purpose.

Empowerment Through Consideration

We are not independent from these lives around us. We need them far more than they need us in order for us to survive. There is nothing more basic or more true than this.

The fundamental principles of healthy human relationships apply to all relationships, including those with life in Nature.

Our connections with Nature thrive in kindness and radiate through considerate actions, approaches, and ways of relating. We are empowered through life in Nature.

We can thrive in modern life alongside Nature. We need to genuinely want kindness with life around us if we wish to progress. These lives are worth it, and they have much to share with each one of us.

Knowing this and truly living it are very different things. We cannot walk away with information alone and expect wisdom without listening. The answers are found through daily time spent with life beyond ourselves: in moments as brief as walking to the car, hours in the yard, sitting on the patio, or observing our neighborhood from our front porch. As we shift our focus outside of our thoughts and relationships with people to be with individual life in Nature, they begin to sense our shift. We need to begin responding differently in situations where their needs are placed before our own.

Be gentle with yourself. Sometimes these moments require time with the individual lives that are Nature. Clarity is not always immediate. What is immediate is the feeling of becoming better. Our body understands before the mind does. It may take time for the brain and thoughts to process what has been experienced. The mind becomes fluent while words fail us. That is okay.

It is up to us to make the time. Value these relationships as personal, and understand that our goals may require the same gentleness. Remain considerate, gracious, and open. Stay mindful of boundaries. Allow other life to "teach" us. Let go of the belief that humans must know everything.

Observe without projecting or assuming. Hold a clear and grounded mind, body, motive, intent, and spirit when among the natural world. Offer respect for boundaries, openness to information, calmness in presence, and love within the heart. The balance comes.

Just like any relationship, when we commit to spending time together and growing in trust and respect, we begin to feel deeply connected. This evolves within us, too.

We know the relationship with Nature is special. Feelings arrive quickly through our senses. The brain catches up later. These connections continue to grow through repeated moments and shared time. Can we not see how much we are worth this?

Our Relationships with Nature Matter

You are empowered. I am empowered.

The interconnectedness found with life in the natural world becomes empowerment rising from our cells—that inner place of health and balance that continues to exist no matter what. People are extraordinary when working together toward goals rooted in appreciation, trust, and respect.

Just outside our doorstep, beyond an open window, among our Houseplants, or within the look we receive from companion Animals' eyes, connection awaits us. When we empower life in Nature for who they are, that same empowerment reverberates back within us, interconnectedly.

Once we accept Nature into our lives as part of our world in a real and meaningful way—where these lives exist in their own right, with needs, intelligence, consciousness, and relationships—we experience deeper connection. We experience moments of awe.

These moments do not happen simply because we decide we need something. We are entering their world, and we must be mindful and grounded enough within ourselves to allow them their own time, just as we desire our own. Slowly, the long-held emptiness within us begins to fill. We are touched by this natural flow. Often, we are deeply moved. We begin to look forward to these moments together again.

When we make time in our daily lives for reflection and connection, we begin to feel compassion, care, friendship, and support—so many positive sensations that the emptiness begins to heal and remain healed. Interconnectedness grows within us and begins to reflect outward through our words, actions, relationships, families, friendships, and perhaps even new connections. This is empowering.

Our senses are relied upon so that the needs of these lives come to us, rather than from us. We sense, and then we know.

Gratitude's Role in Empowerment

The empowerment we exchange with one another not only provides the support we need to feel interconnected—it deepens our interconnectedness itself. This interconnectedness grows within us and throughout the relationships surrounding our homes and communities.

Empowerment strengthens both physical and mental health. Through it, we heal and grow stronger over time—not through the false strength of insecurity that erupts in anger or control, but through calm, healthy, intelligent strength...and kindness.

This may feel like a great deal for some people to take in or even consider, though for many it does not. For those of us who long for these lives to be part of our own lives in real and meaningful ways, it can still be difficult to fully remember and embrace. Knowing what feels right may even leave us momentarily paralyzed.

Life today is complicated and altered in ways that separate us even further from one another and from Nature. Yet we can still find our own balance while embracing balanced relationships with other lives. We can focus our attention on the present—within ourselves and with the individual lives found throughout Nature.

Focus on what matters for all of us and find a balance that works. Doing so may begin to correct many of the wrongs we have created and help us make meaningful progress each day forward while considering the lives around us.

When we acknowledge the inherent worth of life throughout our homes, yards, neighborhoods, and communities—rippling outward through habitats and carried in the winds through the energy of our actions and intentions—Nature responds in gratitude.

Our relationships with Nature matter because they are among the most fundamental and supportive relationships we have. They teach us how to live with soundness of mind and intention. The impact of gratitude on our health and happiness—through the chemical changes it creates within us and the balance it restores to our bodies and minds—is extraordinary. Gratitude is deeply felt. Only the wounded ego struggles against it.

The way we approach life in Nature mirrors our sense of self, even when that self carries pain and suffering. These relationships teach us how to be good people on this Earth.

All life on Earth was created to be here. Our worth is reflected in how we recognize the worth of others. Gratitude does not create inherent worth. Empowerment is never one-sided. Gratitude arises when we embrace the inherent worth that already exists within another.

Mama Squirrel and Lemon

Mama Squirrel was very special to me. She stayed close, though she never sat beside me on the ground the way Lemon did. Instead, she remained in Lightning Tree and ate while we quietly shared one another's company. She was a mother, and I made sure she had what she needed from our small yard—a few unsalted walnuts and water.

She lived in Trees across the street and may have been part of the group of Squirrels that once ran through the yard in single file—an interesting interaction witnessed only this one time by both my husband and I.

Mama Squirrel came every day, usually in the morning before the males arrived. If I was a little late stepping outside at dawn with the feeders and walnuts, she often waited for me. She would sit in the crook of the magnificent branches, eating and watching me refill feeders or adding water to the birdbath.

The morning I found Mama in the street is difficult to revisit.

I was walking Willow when I saw her there, lying in the road with her head crushed. It was Mama for certain. She was on her way back to her Tree. Now she was gone.

I felt sick.

She had been leaving our backyard to return to her den after eating and drinking.

Like people, Animals suffer tragic deaths. These tragedies have shaken me with deep sensations since my childhood. I could sense the fear and the moment before she was struck by one of the cars that sped far too quickly down our residential street.

Once I walked past her, the feeling consumed me. I felt responsible in some way. She had been leaving our yard. Regardless, I would never see her again.

It is very difficult for me to see Animals in the road who have suffered immediate or prolonged, painful deaths. I sense the fear, the pain, the loss—the terrible fear.

I cut our walk short because I feared she would continue to be run over without respect or regard for her beautiful life.

Back in front of our home, Willow and I stood quietly. I wondered what to do. I worried what people might think if they saw me doing what I felt so deeply called to do.

Then I decided.

I would stand by who I am and by what I believe and know. Willow stayed inside while I returned with a towel.

I knew Mama Squirrel as an individual. She was magnificent. She cared for her babies and returned to our yard again and again. She kept mostly to herself, yet she felt like family to me.

I gently picked her up in the soft towel without looking around to see whether anyone was watching. I held her lifeless body, still warm from having just died.

She appeared to be pregnant and, given the season, that seemed possible. She was heavy.

I carried her into the backyard where I had first come to know her. I buried her there, near little Peanut, and Flowers eventually grew from Mother Earth above her.

I cried because I felt the loss of my sweet friend.

Two days later, Lemon was hit as well.

Only his ears were recognizable. I knew Lemon by his ears.

He died from a car, too. It was early in the morning, possibly by the same person who sped down our street every single day.

Lemon died to the left of our home. Mama died to the right.

These are losses I still carry with me to this day. That week, I did what I could to help them continue on their journey.

In these two moments—my final moment with Mama Squirrel, and then with Lemon—I no longer cared what others thought of me.

Our Silent Breath

Love holds a sacred place.

Life around us continues to hold a sacred place within me because of individuals like Mama Squirrel and Lemon. In moments of meditation, I am often with them again.

I knew their worth independent of me. I embraced their inherent worth because I knew them as individuals in their own right—because of who they were in the moments we shared and because those moments changed me for the better.

Through them, I began to see worth within myself in places where it had once been underdeveloped or perhaps lost. Their worth empowered me in ways that remain with me still.

Because I came to know them, and because I offered what was needed in response to what they expressed through their lives, the need for approval from others slowly loosened its hold on me.

In warm brown eyes, shifting ears, and the joy and freedom radiating from fur and whiskers, I felt empowerment move back through me and remain within me. My inspiration deepened. I was changed. Our interconnected senses—and my Nature within—remain intact, binding us always, silent and present with every breath.

Twenty-Five
Knowing When We Are "Off"

Sense the Balance

Play is so necessary. Individuals in the natural world play in low-stress environments. Around and within ourselves, we need the same. Laughter heals. Remaining present eases the brain's functions from lives of constant stimulus processing. We feel better. But during those periods and instances when we are feeling "off" or not right, we can sit among our brothers and sisters in the natural world. Our bodies learn to acclimate more readily and easily. There is no pressure and, like Dogs' joy for a walk, we feel the same. We return to what feels like a much grander home.

Balance is a calm, content knowing and focus that eases our heart and quiets the brain's activity. When we feel balanced, essential communication within our bodies that may have been neglected now has space to come alive. Outwardly, balance is confidence; a light or "lightness." The systems that function within our physical, mental, and emotional selves operate as designed, as created and evolved with Nature, not removed from Nature.

Balance is everyone doing their part. Working together and living together—finding solutions together. Balance is so powerful that others are drawn to those steady in balance without even realizing it. Nature pulls us in, too.

In all our attempts, strategies, and abilities to remain balanced, we live in a world that will toss us around every now and then. Nature supports us so that we can be as we are meant to be, naturally, in calm waters. Without judgment, as unique as we are, and as varied and different as we each can be—in balance.

Our needs for balance may vary, but the need for balance is the same for all of us, including life in Nature. This is Nature. However, for us, returning to balance when we drop back into old ruts of who we were, what we have done, or what others have told us to be can seem impossible. But nothing is impossible. Taking moments and committing to balance allows us to become more complete with both the modern world and the natural world. Offer this same respect to all life around us. There is unlimited potential in having, respecting, and appreciating our boundaries and the boundaries of others. There is no hierarchy here with Nature; only balance stabilized by boundaries.

The first step is to come into relationship with balance itself. To do this, we accept that stability is necessary and deserving of attention. We understand that, although we might not understand everything about our interconnectedness with life in Nature or our need for their kind relationships, we are entering their world, too. Trust in the fact that we need them in our lives in real ways that allow them to thrive as they need. We acknowledge that, for each one of us to thrive, we need to offer this to life in Nature. When internalized as understanding, stability begins to set in.

When we are steady within ourselves—when we are confident and supportive of who we are; when we know what we feel and, perhaps, why we feel what we feel—we become comfortable with this, too. We are comfortable with who we are and, in this confidence of balance, we wish the same for others. We want to be better because we feel better—inspired. This calm stability is felt throughout our body, throughout our physical self. This calmness is also felt by those around us, including life in Nature.

The natural world and the individuals of Nature are much more in tune with relevant sensations and senses. Understanding this and the impact our state and presence have on those around us, because of the direct impact on ourselves and others, is important. We become what we are within.

Since negativity can cause physical harm, affecting anyone in its path and those close enough to feel its impact, balance becomes the foundation of our state of being in a very real and outwardly directed way. We become active in caring for the core of how we are across all aspects of self. These aspects—depending on our condition—work together to support us: our physical, mental, emotional, intellectual, and perhaps spiritual and mindful states, forming our whole self. We allow balance to be present within our bodies. This becomes our greatest way of finding our center and supporting ourselves in any and all situations.

We need time with the natural world around where we live to have our moments together, quietly. This time fuels us to honor our personal strengths and our very real gifts without human-induced ignorance and judgment of self.

We are more like Nature and, therefore, they, too, are aware that our intentions are to honor their worlds. Here, in our calm, reciprocation can flourish.

Quiet, Mindfully Expressed

Know the power of the energy behind words and how they trigger our bodies. Continue to refer to life in Nature accurately, and not as an "it" or "things." Become very aware of personal representation. We are what we radiate, including our words. We might slip for years, reverting to toxic thoughts and negative words, but our intent is different. This is the difference: Our energy released through our actions remains supportive, helping us on days when we feel unsure.

Having cause to stop, enter within through our mind, and listen to ourselves—our inner voice, our knowing, and our sensations—supports our awareness. Doing this over time becomes easier and faster.

Some of us may already do very well in reaching this state due to our upbringing, experiences, and current practices. It is a state of knowing ourselves. It is almost like meditation in the moment, but it seems to move through a door into the realm of information and knowing—entering the silence, the quiet.

In this space, sense and feel gratitude. Nurture it. Create the inner environment for appreciation. Holding this dear to us alone matters greatly. Our cells drink in gratitude. In our gratefulness, remain aware that the lives of the natural world matter greatly. They are part of the soil of our personal strength here on Earth and all around us. We are gifted with this by default in our inner awareness. Gratitude triggers something so powerful within us when walking with Nature.

Remain aware of our physical, mental, and emotional parts of self. We will be alerted when something needs addressing. Pay attention and offer quiet with Nature so we have what is required for our balance. We can be what we need of ourselves and find the same with others.

Adaptability is necessary for life. To have the ability to adapt, remain open, and learn is vital for survival. Learn that perhaps what we thought or believed about other life around us may not have captured all relevant information. Perhaps the most accurate and positive response to the expressions of their lives was lost on us. We need to be okay to feel and feel for others, especially life throughout the natural world. For those of us who are highly empathic, this may be our core strength. No one can create it for us or tell us how to be it. We are our own source of strength, supported naturally with Nature.

Balance is working with life all around us. Our within is no different. We feel it when we are "off." When we feel good, we take it for granted, too. Respecting these individuals in their health, happiness, and balance reciprocates the same within our own mental, physical, and emotional states. We need these lives within our lives, in our awareness, and at the forefront of mindfulness. In our quiet, within, it is we who must establish basic trust in their well-being as part of our way of life...a personal path appears before us.

Remain aware of this support in bad times, but also feel it in the good—again, gratitude. Gratitude, and taking moments quietly with Nature and quietly within, does something miraculous to all areas of our health—mental, physical, emotional, supportive of any part of kindness in the spiritual and intellectual. Gratitude is like the icing on the cake that creates a clear state regardless of any perceived flaws and imperfections.

Life in Nature does express gratitude. For us, being just as considerate is natural and necessary.

Transform Our Focus

Our thoughts, and what we decide to focus on in situations, experiences, people, places, interactions, hopes, and dreams, swirl within the brilliant life vibrating throughout the natural world. Not all people take action to be considerate and positive behind their words. Change needs to happen, and remaining positive helps guide us like a light. There will always be negativity pulling at our focus, along with hurt. Change our focus to include all that does not.

What and who we choose to place within our field of focus—when it comes to setting our intentions, focusing on what matters to us, and who we want to be and represent in the world—affects us. The corresponding thoughts change our body through chemical reactions within. Our thoughts are triggered by our senses and experiences, and those senses burst into life come through immediate emotion, which may lead to other emotions or to the suppression of emotion—all of it flowing within us as information.

With life in Nature, our focus turns toward an environment for the positive, the supportive, and the caring. This might mean simply leaving an individual or group of individuals in the natural world alone so that they can simply *be*. The act alone heals us and offers them the same grace. The act of kindness or grace is one of powerful confidence.

Anger Is Fear

We—humanity as a collective—do not feel right. We do not feel well with the ease that comes from within. We feel it in our gut and in all of the neurotransmitters present, telling our bodies to begin processing this information. Sometimes these sensations of anger grow from fear. Fear feels awful and nauseating when mixed with anxiety. Some release these feelings onto others, and many turn to anger when fear creates cracks within.

Still, we do not feel right. Anger is hot, and it burns us from within. Outwardly, we can and do spew flames and daggers. We need to release it from within. Through words, our release of intent can become twisted, altered, and false. Anger is the human-centric interpretation—in all of its varied forms and expressions—and often it is rooted in fear. Fear becomes toxic unless we wrap our calm around it, within it, know it, learn from it, and release it so it can dissipate cleanly. We feel the difference.

When we allow negativity to be injected into us by events, people, situations, or our own mindset, self-doubt and our great talent for beating ourselves up with negative thoughts take over our inner landscape. Balance becomes overrun and shrinks within us. We feel it within. This, too, is felt by those around us in Nature and by any person sensitive to the subtleties of others. The off-balance can exit our mental, physical, and mindful states as anger, either through words, feelings, or actions.

When we know ourselves very well, we know what is ours—how, what, and why we feel what we feel—and what emotions belong to others. Do not let another's anger become your own when it is not yours to have in the first place. In this moment of feeling "off," turn your focus to boundaries.

Anger sticks and can pierce through or remain draped over us. For those who are sensitive, we can learn to see through anger and recognize the fear; know if it is ours or if it belongs to another. Here, we can see the fear clearly. Fear seems to linger as a scent. We can learn to distinguish another's experience from our own and return it to the source through boundary knowledge within our quiet, calm mind. We can clear the air of fear that is not ours, and it is possible to allow another to be responsible, perhaps accountable, for their own fear.

Even in high-stress or busy moments and days, with anger flowing all around us, we can still feel balance. Balance helps extinguish the fire of anger and opens the light-filled window that releases fear. We can choose to stand in this light—a sense of protection. Stopping to sense where we are emotionally in the moment helps bring this awareness forward.

Balance remains intact when we appreciate life throughout Nature for all they are because we do the same for ourselves in the mirror. Nature will stand with us. Anger used for our protection when we are not treated well can help us adjust so we can grow. Then release it. Listen to within; listen to Nature.

Negativity creates more negativity very easily. It is the easy way. It takes a strong, intelligent, and wise person not to harbor it, but to turn negativity into learning and create something positive from it—to release it as positivity that dissipates all negativity. To be more like Nature and life in Nature is strength and brilliant wisdom. Nature operates on fear. It is we who contort actions or reactions into anger. Confidence needs no façade. Anger is fear.

The Red Flag of Guilt

There is something very human about guilt. Guilt is not an enjoyable thing to feel or carry around within our bodies, minds, and thoughts. It does not take much to make a wrongdoing right when we communicate. But some people will use guilt to manipulate others. Given how poorly we often talk with, rather than to, one another productively, it is no wonder that even when we try, interactions can still leave us reeking of pain and hurt. Abused human words pierce deeply, and some are unwilling to accept responsibility in moments of wrong. For some of us, we take that negative energy that belongs to another, turn it around, and force it within us as guilt, even when we know in our hearts we have done nothing deserving of the guilt.

Some are so unhappy that, in that state, there is a palpable elation in releasing their pain onto others through twisted forms of manipulation. These are the bullies. These are the ones who smile to our faces and break trust behind our backs. Those who refuse to take responsibility become so disconnected from grounding support and interconnectedness that, even in their lack of clarity about themselves, their attacks only deepen their own original source of pain, sadness, and despair.

Many of us are very good at feeling the guilt of others and confusing it with our own. So, we search within our world for an action we have taken or words we may have said that we should feel bad about, when, in fact, we have not done wrong. Guilt tastes a lot like control—someone trying to control. But we can become masters of our own health while remaining sensitive to the experiences of others. We can learn how to sense guilt, stop ourselves, and know whether it is ours.

In those moments when the actions of others are clearly not rooted in kindness or invested in the good of another, guilt worthy of apology hangs heavily in the air. For some of us, as the target of another's guilt, we breathe it from the air, feel it, and sometimes become it. We are the ones who end up apologizing for so much, so often. When we live in these situations long enough and grow numb to the confusion, we can no longer identify the source of why we feel guilty. We have learned to take it in as our own. Then it can turn into shame. Shame is toxic.

It can become very clear that someone needs to take accountability and apologize, yet those who are irresponsible, emotionally stunted, or lacking emotional maturity altogether may instead spew guilt onto others in attempts to feel better themselves, in some disturbed way.

There is a health in guilt, too. Its presence raises a flag that something needs to be addressed because no one is perfect. We only need to respond to what has been flagged within us, return it to its proper place as future guidance, and act accordingly to right a situation where needed so that we may stand firmly and easily in our lives.

Those of us who are sensitive to guilt and the negativity lingering within relationship connections may grow accustomed to taking accountability because no one else will. But it is not ours to carry within. Go to life in Nature near our homes and communities to feel the clean, clear air of life that seeks support and balance. Reciprocate this and hold it close within. Here, we are seen and heard without a word spoken.

We Can Choose to Evolve

Being among those around us who seek balance and support can pull us from the confusion of people at times and provide clarity. Even hurtful people have been hurt, and they, too, can choose to place emphasis on the good of life throughout Nature. Healing begins by breathing Nature's intelligence in so that this same breath flows through our words and intentions toward the people in our lives.

Our support and gentle guidance are interconnected with the lives outside our doors and around our homes—the individuals who see us and know us within our home environments. Life in Nature is aware of our treatment of other members within their community. Support comes back to us fully and offers the grounding we need to return to balance and health within, especially when we are feeling "off."

Going forward, we can choose who to trust and who not to trust, just as life in Nature does. Sometimes we are simply not compatible with someone at a certain point in our lives, and it is that simple; there is no resentment or ill will, only a desire for what is good for everyone. At times, others can manipulate our vulnerability; we know these individuals well. Trust is earned when good human motives become equally well known—felt, sensed, and present in the air. Nature understands this, and we can too.

We do have a presence, and much of that presence comes from the energy we emit to those around us. Become acquainted with our own emotions and feelings so that we may become more aware of our words, intentions, and actions, and how they radiate out into the world—to other people and to life in Nature.

If we decide that our unhappiness and struggles are out of our control, or that others are responsible for making us happy, or we refuse to accept responsibility for what we have allowed to transpire in our relationships, toxicity is inflicted upon our Nature within—in parts of our physical, emotional, intellectual, perhaps spiritual, and mindful conditions—and radiates outward from us. Our thoughts, actions, emotions, and senses trigger our brain and body functions. We really do become that which we think—our cells know. Life in Nature is keenly aware of this, just as they are aware of our goodness. Whatever we exude permeates our pores. Our desire for calm and balance is expressed clearly to these individuals through our intentions, whether spoken or unacknowledged within ourselves.

We can choose either to evolve and grow positively and in health, or to continue living in ways that are not working for us. When we struggle, when we have days where we do not feel right—perhaps "off" balance or off-center—our actions must change if we are to change, beginning with our thoughts. Focus on kind mindfulness with the world around us. Trust in the good that is created when we wish no harm for another and only want what is best for them. Life in Nature also responds to this positive shift.

Twenty-Six

Remaining Open Is Our Powerhouse

Inner Strength Is Fueled by Confidence and Respect

Humanity is in a position of individual strength. This is ours to claim or to lose. It arises from the confidence and willingness to remain open to the life around us. In the face of struggle, we can return to a sense of belonging with the lives around us, and this begins with our intelligence.

This is not of thought, but something felt within through our senses and the emotions they trigger. In time, our inner home becomes stronger than any struggle. How we handle ourselves and our connection to particular situations going forward is our responsibility. This is the knowing of self—providing our sense of self with what it needs as emotionally mature individuals.

When we develop and hold dear our relationships with these lives in Nature, our powerhouse of remaining open offers unconditional support. We not only learn about the amazing individuals who call our yards and communities home, but we also learn about ourselves. Nature mirrors us in many ways and offers what we need, even when we may not know it ourselves in the moment.

Our greatest asset is our supreme inner intelligence and presence in this world—one that impacts our relationships, our actions, and our ability to learn and grow. What we sense and know through ourselves and through who we are within meets the world around us, especially the highly sensitive lives throughout Nature. Nature quietly feeds our inner intelligence through these sensing relationships. We come to know ourselves and others by listening to this information. Inner intelligence takes many forms.

Our empathy and our empathic sense of knowing and understanding ourselves and others extend from our emotional intelligence and this inner perception. Emotional intelligence balances us, calms us, guides us, and allows us to understand life around us, especially life throughout the natural world. This intelligence connects us with all life in Nature because interconnectedness flows here. It is our ability to become our own greatest support and confidant, knowing why and when we feel what we feel.

Emotional intelligence equips us to care for ourselves and to support our boundaries with great strength when we draw inward and hold fast to our inner intelligence. We know what our triggers may be and what we need in order to feel better. We know ourselves well without needing to declare it. We hold this close while living in a world over which we have much less control. With our feet planted firmly on the Earth, we are supported in who and what we are, where we want to go, and who we want to be when guided by kindness and actions that create the positive. Our intentions matter, and they matter to our sense of self as well. Trust within radiates trust outward.

When we trust ourselves, our self-respect blooms and grows. We become like life throughout Nature, unencumbered by the false hopes and insecurities of others. We are no longer weighed down or confused by who, how, or what we "should" be. Instead, we stand, exist, and coexist. This is life with Nature. Through trust, we develop deep and unwavering respect that sustains who we are within our cells and at our deepest level of interconnection.

Here is where we connect ourselves with life in Nature. When we feel lost in our own emotional agility and in what it means to be with individuals in Nature, we are shown the way, and we learn. This is the voice of Nature.

When we finally understand the value of our emotions and how they connect us with ourselves and with others, including life in Nature, we begin to appreciate the intelligent strength of balance. We discover the strength of being fully present in our inner state. We become grounded. In finding camaraderie with our brothers and sisters around us—with Trees, Dragonflies, and Swallows—we see ourselves and come to know ourselves. Through knowing them, we become more open to knowing ourselves.

In this clarity, we experience moments that inspire and open our eyes. These moments are personal. They are ours, and not something any other person can create for us. They are meant for us, offered by another in the moment and grounded close to Earth. Though often misunderstood, these lives remain open to us. First, we need to be worthy of the trust we seek.

A Genuine and Influential Trailblazer

We cannot expect from others what we cannot give, nor what we cannot give to ourselves. Nature is aware of this.

As we learn from and with life in Nature, these relationships bloom and expand. Perhaps, in time, these lives come to see us as one of them—uniquely individual, with our own boundaries, while honoring their individuality, boundaries, and need for space as well.

Reciprocation happens through interconnectedness. In our powerhouse of remaining open and embracing what, why, and how we are, the brightness on the other side is always better than remaining in the muck of uncertainty and unclarity.

Thought is different from feeling. Here, sensing and emotion come first. We are our greatest support. We owe this to ourselves. Our actions grow along with us, and they, too, evolve as we grow with life in Nature.

Facades of personality fall away in the presence of life in Nature. Each one of us is our own trailblazer. It is up to us to blaze our trail and become the powerhouse we are, intertwined and inseparable from life in Nature. We *are* interconnectedness.

From today and each day forward, we offer life in Nature space—both literally and deeply. This transcends our sense of care for these lives. Holding affection and appreciation for these individuals and natural communities within our breath develops these relationships in ways that strengthen all of our relationships, including our human relationships.

When we intrinsically embrace living in alignment with our interconnectedness, we act responsibly regardless of sentiment or passing feelings. This is not a projection of human states onto other life. That would never do justice to them. We cannot claim to understand them completely. Instead, we walk in tandem with the life around us because it is the right way to walk according to what we believe within our hearts.

We are each unique individuals, and it is not our burden to carry another's flame. Our character matters. We need to extend this same respect to the lives with whom we share this planet Earth.

We are magnificent. We are our own powerhouse within. Remain cognizant of the life around us—kindly, compassionately, steadfast in openness, and always offering space for the uniqueness of each individual, especially life in Nature, as our roots expand to all life around us.

Our relationships with individual life throughout the natural world run within our blood—literally, and through the positive experiences and memories we build. These relationships flow through us physically. When we are trustworthy because we honor boundaries, appreciate their needs, and always provide space, even when we do not understand everything, these lives become supportive, calm, and balanced, and they want the same for us, interconnectedly.

Respect is earned when respect is honored and given. Deep, genuine respect cannot exist without trust. Nurture trust within yourself and with life in Nature to restore it and uphold it. From trust—because we are trustworthy—we can breathe out and breathe in what it means to respect these lives.

Jazzy-Like

Remaining open to the life around us—who they are, how they live, and their intelligent skills, talents, abilities, and emotions—builds interconnectedness within us.

Remaining open to ourselves is equally imperative. Remain open to our needs, cares, and concerns, and support them through self-trust.

Accepting who we are and honoring our boundaries, in a world that often does not, gives us the strength to stand present in our own state of being. When we seek calm and balance, life in Nature supports us because this is what they know unconditionally. We become humble when we remain open.

Part of remaining open is ending our dismissal of openness and our dismissal of the life around us, or belittling these lives and any attraction to them as childish.

While centering and balancing one spring morning, I reflected on the fact that I had never truly seen one Opossum in Nature who was not killed by a car. Perhaps once I saw one running off a road, but it was little more than a glimpse of a small body and tail.

The next evening, I happened to look out back just as she came around the shed. There she was—large Opossum, whom I call Beauty. She walked across our deck and over to the right, beneath Evergreens where young Squirrels play. She found something on the ground and began to eat. What a remarkable sense of smell. Truly amazing. Then, she left as quickly as she came.

I was so grateful that I looked out back when I did, or I would have missed her completely—and missed her message meant for me alone.

Our former neighbors had a vegetable garden, so I suspect these friends had been visiting for some time. The house behind us offered a great deal of space where they can live undisturbed.

Another morning, I happened to look outside as I considered feeding Birds earlier than usual. It was around 5:30 a.m. I opened the curtain to the backyard, and there she was—very large, beautiful Skunk whom I call Jazz.

She came around from the shed and paused in what she was doing. I believe she noticed the curtain opening. I had been aware of her lingering scent near the shed, so I was not too surprised. She stayed by the shed and ate small seeds that had fallen to the ground the previous day. She was very pretty.

I think she heard Squirrels running across our roof, or perhaps something else, because she quickly ran to the back of the shed with her tail raised in protection. She was very jazzy-like; very, very beautiful.

A few minutes later, I went outside and could still smell the familiar scent near the shed. Skunks leave a lingering odor even when nothing has been sprayed. This occurred less than a week after seeing Beauty. I felt very fortunate.

Since then, I have often been out back in the dark with Jazz. She knows me now, as does her Skunk companion. She stays by the shed doing her thing, and I remain on the other side of the yard doing mine. I do not stare at her. I simply let her be.

She came every morning, and we respected one another's space with confidence. Jazz was here before me, after all, and she is part of where I live. This is her world. I give her space, I respect her boundaries, and she reciprocates the kindness.

Twenty-Seven

Interconnectedness Comes Naturally

The Joy in Basic Needs

We have a need deep within each one of us that reaches and searches, extending beyond our physical bodies, our brains, and our minds, seeking to interconnect with life beyond all that we are as people. This is not to say that healthy human relationships are not important. Nor is it to say that any connection or devotion one might have, or wish to find, within a particular group of faith or spiritual beliefs is not important. These are very personal and very specific connections each one of us might choose or feel drawn to in our lives.

Our interconnectedness with life in Nature is a basic need, like food, water, shelter, and space. This need is as natural and just as important. We need life in Nature within each one of us by respecting these relationships all around us.

Any demented or distorted demise inflicted upon life in Nature is a great deterrent with very real links to our physical, mental, and emotional health. Keeping our bodies going and safe matters. Keeping our physical, mental, and emotional selves healthy is just as important. We will never live and breathe in optimal health and balance without our interconnectedness, and this can never be complete without life found throughout the natural world.

Realizing the magnitude of basic need fulfillment is crucial here. So many in this world are not as fortunate and struggle to obtain these needs. By focusing on what matters at the deepest part of our survival and existence, we discover very real and unwavering joy—joy in the moment. If we need more joy and happiness in our lives, perhaps it is time to turn to life in Nature as the one in need, so that we might appreciate their kindness, support, and camaraderie.

Finding joy with life in Nature reciprocates and reverberates back within us. They support the other areas of our lives that need kindness and love. We are supported in every part of our life that matters deeply to us when we considerately support kindness as a whole.

Creating moments of joy by being with Nature has nothing to do with a need to enter another's world, far away from our own, or with trespassing in areas that other life needs for solitude. Life in Nature has needs and boundaries just as we do. What motivates us most, what matters most at the end of the day, or when we are confronted and have no choice but to face it, is this interconnectedness. Like our other basic needs, happiness and joy abound here.

Our interconnectedness with the natural world reflects our self-respect, and that is mirrored through our respect for Nature—not for resources or for what they might provide in terms of money (something human-made and not of Nature in any way). Within our interconnectedness lives our motivation for what matters most to us.

We still have an internal light flickering softly within us that wants to do well for the natural world—a light that cares deeply. Happiness and joy expand into purpose. Purpose is our need to create a focus on good in our lives and for the sake of all life on planet Earth.

Follow the guidance offered by life in Nature, both where we live and within our communities. These places need us. The far-off places also need us to remember where we live and what matters most.

The Light of Gratitude Flickering Softly Within

We are openhearted to appreciation—gratitude—in simple moments throughout our day, if we are fortunate. For some, we dedicate time, in quiet and within, to express gratitude. This is part of our health. With life in Nature, appreciation begins to seed, grow, bloom, and spread.

We are not the keepers of gratitude. Gratitude is prevalent in Nature. Perhaps this is where we received it, too, once upon a time, and why so many of us feel so far removed.

Gratitude comes in many ways. It can be felt, witnessed, and sensed without words. When sensed with life in Nature, by taking time to hear and learn from these lives all around us— around our homes and communities—gratitude spreads like the wildflowers of native lands throughout our bodies, our lives, and our relationships, including those with other people.

This is a beautiful strength of self that resonates from a realistic and positive relationship with the natural life around us. There is no need to attempt manipulation. Negative or ill-willed motives will not take root here and will never survive. This is not the way of life in Nature.

Welcoming gratitude interconnectedly creates and builds. This is interconnectedness in action, with and through life in Nature. We, too, can become part of this presence and intention. Gratitude itself is welcoming. To know it, we must be it. To be it, we must be grateful.

A Return to Balance

When we are with life in Nature, when we are with individuals in special places we protect and call home, something happens within us. When we take time to quiet our minds, offer space, and allow for the needs of others who are nonhuman, our actions are noticed. We are noticed, even if mostly hidden from us. *We are noticed, still.*

As our intent in these special places becomes considerate, dependable, and clear, we begin to build a bridge between ourselves and another, regardless of who they might be, and something special begins to happen...

We are known as individuals. We become known through these kind actions and are seen for who we are within. In our sacred space of home, these individuals begin to balance us in ways we may never have expected. The calm and living force of our yards and communities root deeply within us, too.

These are the places where we raise our young and care for our needs. This fertile ground of home offers the same consideration for another to be with us, as they need, for this is their home, too. And when we are grateful in our interconnectedness, we want what is best for them as well.

Through our trusting interactions of mindful presence and inner acknowledgment, our senses are enlivened and activated. Our emotions are realigned, and our thoughts begin healing within each one of us in our own ways through our experiences with these lives. Balance returns to us, within us, and reaches our homes.

This is independent of other people or those with whom we share our space, and it is special and unique to each one of us: personal relationships with individual life in Nature. As others focus on their own process with life in Nature, a common thread of understanding begins to connect us through these lives around us—balancing Nature.

Balance is very personal and very specific to each one of us as individuals. So much so that what has been restored, maintained, or provided for our balance is to be shared only with those whom we trust deeply, and only when needed. This calming breath is for us to inhale deeply. We do not need to explain it away or justify it in any way. Our state of balance is deeply personal and requires no validation.

Life in Nature supports our balance in profound ways. We share when we feel or sense that we should. The rest is our own personal relationship, openly supported and safely held by all life throughout Nature as part of our uniquely valued interconnectedness. Positive, considerate, meaningful, and supportive, it is built from Earth up through us first.

Hummingbird Said, "What We 'See' is Perhaps Our Own Difficulty"

We are all in need of interconnectedness as a *right* shared within our community of life on this magnificent, uniquely special, and beautifully remarkable Earth. We have endless support all around us in our neighborhoods and communities, for endless life calls these places home, too.

It is here that we need to develop interconnectedness. It is here that these lives have chosen to share space. It is here that we can look up to see the clouds and breathe in the air, just as all life around us does.

I was warming some green tea during a small break from work when I took a moment to look onto our patio and backyard. It was turning chilly, with cooler days in a spot that receives only afternoon sunlight—if the sun shines at all from behind gray skies. One little Bird was eating what was left in a dish of food.

They seem to know when I replenish the food and switch the Hummingbird feeders to the warm one I keep on reserve inside. I have been doing this for years. It is very much appreciated, from what I can sense. Hummingbirds come down from high inside Trees and look into my eyes for a moment or two before leaving.

One day, I took only a few minutes to change my shoes and check the seed levels. The small cup for Chickadees and Nuthatches was empty, despite my filling it earlier that morning. I have finally found harmony on the patio. Harmony and balance are important to me in the places I share with these lives and remarkable individuals, some passing through on migration. I can tell by their cautiousness. Others know me now, too.

The Chickadee and Nuthatch cup hangs in a wind chime beside Lavender in a tall planter, preventing Squirrel interest. The Squirrels now focus on their five small feeding areas placed in different locations to provide space and an easy rinse of water from the watering can each late afternoon.

The Hummingbird feeder stands separately against the house, away from a window and a good distance from tall Grasses planted in a large pot that sway frequently. Two small feeder plates are placed out for Birds. One additional tiny dish is set in a separate area for Millet only. Juncos enjoy this, along with Song Sparrows and our now-comfortable Spotted Towhee.

This arrangement gives everyone space. The two dishes hold hulled Sunflower pieces for fast eating. Everyone likes this seed, and it serves as a fallback for many who now gravitate to their own relaxed spots.

That day, though, in the chilly afternoon air, I sensed something different. Hummingbird began with his or her high-pitched sound. I only learned this sound last year. I thought it was a warning, and perhaps it is. But that day was different.

Crow—always stationed around the area, with two or three individuals strategically placed, true to Crow law—began calling as well. Two Squirrels arrived. More Birds came, too.

It may have seemed that some were simply responding to the sounds and activity of others, but I believe differently based on what I have witnessed over the years. These Squirrels are part of a group that dens in tall Evergreen—the same Evergreen where young Crows stayed while their family was out and about, always within earshot.

Crow babies have been left with me on the ground in my yard in the past, along with baby Robins and baby Squirrels. Squirrels and Crows work together to help one another. I believe the Hummingbirds do, too.

What appears to be chasing, I have witnessed as interactions between males and females when I once thought they were fighting. What also appears to be chasing may be one Hummingbird showing another a food location in their own way, since they move together and rest in tall Trees. When I reflect on these encounters in familiar places, my default interpretation tends to focus on territory and competition. But when I set aside analysis and allow my senses to meet the experience more directly, I sense cooperation—support, communication, and care. Across species, within specific relationships, and as part of a broader community, there is often a different kind of exchange than what first appears. Even in moments centered on survival or caring for their young, there can be an underlying rhythm of awareness, balance, and responsiveness—an intelligence working together for the good of the whole.

Be in Nature where it matters most and where we can be most often—where these lives choose to share space. Our inspiration is suffocating along with these brilliant lives all around us.

We are part of this interconnected equation that is life on this planet. Our physical bodies, our thoughtful intelligence and minds, and our spirit that enlivens and supports our compassionate and powerful beliefs are all aspects of ourselves that we need not only to thrive, but to do so in health and happiness.

Life in Nature cannot continue under our current weight of intolerance unless we, as individuals, embrace these relationships. With Nature's reciprocity and support, we begin to see ourselves in our connections. In turn, we are embraced, and we heal.

We offer the same in return. In our authentic desire to earn Nature's trust and build healthy relationships with these lives all around us, we honor our very basic need for interconnectedness. We evolve and adapt to the unifying law of respect.

Our futures can include this embrace and learn from these uniquely intelligent lives that are Nature. We can learn and grow from the lives around us. We can evolve with this strength, and evolve together.

Twenty-Eight
We Are in This Together

We Remain as Guests

Our relationships with life in Nature matter greatly to our support, balance, health, and happiness. Without them, we struggle through our deeply neglected need for interconnectedness with life in the natural world. With these lives, we find unconditional support for our own self-trust and inspiration, the roots of which can be found only with life in Nature. It is a support very different from our human relationships—a support that matters on all levels of our being and believing.

There is no need to feel as though we must impress life around us in Nature. We can have personal interests and hobbies as the people we are, while also shifting to a place of embrace with life in Nature.

In many places in the natural world, we remain as guests. In this accepted mental position of good and positive intent, we offer the respected space that ignites the interest and support of life in Nature.

Mindfulness goes deep and far with life in Nature. When we do this, we begin to maintain the same awareness in our homes and yards, neighborhoods and communities. Outside our doors, we meet on common ground with overlapping beauty. These are our places of escape and grounding—places of relating, relaxing, and interconnecting together with these lives of Nature.

In learning to be who we are with Nature, we learn what it means to be kind when so many are not. With Nature's trust, and by taking time and allowing time, they will and do have our back. Once we are truly trustworthy, we become part of a world of confidence that exists throughout Nature. We feel it within, reciprocating back from these lives throughout our days, and we experience moments of insight into where we need to love and how we need to heal.

Nature can show us, but we need to be kind. We need these lives more than ever—more than they have ever needed us.

We Have Something to Learn

We fail, or even refuse, to understand that we have much to learn from the natural world around us in our communities. The same happens among people, too.

It is no surprise that the need to feel superior is so easily welcomed within personhood when, in fact, those needs are rooted in insecurities and fear. We need to change this view on many levels because these lives will not wait for us to do so. The planet, operating as a whole community of life—through these lives for which we have steadfastly decided we want no part—will find a way to balance with or without our cooperation.

If we are part of this planet and share the same makeup as the rest of life on Earth, it seems only logical that we have much more in common than we do not. Know them as they are instead.

We have some catching up to do. We have much to learn about how these ways relate to human connectedness. Perhaps our strengths and challenges with other people are connected to patterns of dismissal we carry within us. Even strangers—each with hearts seeking the same sense of connection, shaped by similar vulnerabilities and longing for support and kindness— are worthy of our attention and time.

Lumping any affection, devotion, or aspect of caring for life in Nature into something unworthy of adult time only adds to our current state of decay, unhappiness, and struggle. Not caring has become second nature, but it is not Nature. We need to regroup and begin to feel and learn from our comrades living all around us throughout Nature.

We still have so much to learn about care, love, devotion, support, kindness, family, trust, and respect from these lives in Nature, regardless of where we live. They matter because they have their own right to matter. They matter because we are a mirror of these lives, and for too long, we have discarded and refused to see that the mirror works both ways. What we see in them, and what we do to them, reflects back to us.

In this reflection, are we offering no credit to so much life and so many systems and communities—individuals who matter—simply because they do not fit into a preconceived box? Boxes do not exist naturally. Communities and environments do. Kindness does. Kindness matters more than much of what we assign monetary or other superficial value. This is where real, secure, and confident power resides.

We have much to learn in areas that seem simple to comprehend and embody. We are the ones who overcomplicate the process that has worked so well.

Only through our honest intent of kindness, trust, and respect might we find all that we are able and desire to learn, and live fully in the present with health and happiness. It is not all about us. It is, however, up to us.

A Fresh, Deep Breath of Consideration

Extending deep consideration to far-off places as the homes of other life has a direct impact on our health, our bodies, and our happiness. We can begin in the personal and private spaces of our own yards and homes, focusing on our communities and what these places truly mean to us.

Let us embrace these spaces as special—our sanctuary, our refuge, and a source of profound meaning. This is where we live. Why would we not want these places to feel deeply meaningful, restorative, and alive with connection?

In our consideration and appreciation for these lives, we begin to breathe in insight and kinship, moving from simple observation to deeply felt sensations. We learn about these individuals, and we learn about ourselves.

In their world, without our human voices breaking through our moments, we notice that they, too, experience gratitude and joy, love and sorrow, loss and play, and support among one another and across species. It only makes sense that this deeply connected need lives within us as well.

We witness their contemplative moments as they also watch, observe, sit quietly, and face both sunrise and sunset. In this space, we may be present and aware enough to sense that perhaps something else is at play—something more profound and universally united with all life on Earth.

I have watched Birds sitting quietly in Trees at the end of the day, all facing the sunset. Chimps have been known to leave stones in special places and remain together in quiet, in what appears to be contemplative thought. If we understand this deeply, is it too far-fetched to consider that the same sense is found in other life?

Something deeper happens when Bird stops, moves close, and stares into my eyes for many moments that can only be described as moving, profound, and very personal. I listen to Nature.

In a world focused on looking out for oneself, many of us have forgotten—or perhaps never considered—the essential good of the whole. There is balance in caring for ourselves, our needs, our families' needs, our yards, and our communities. But selfishness rots the roots and does not reach far at all.

Healthy boundaries provide what we need. Beyond that, the rest is about learning to live considerately as part of a greater whole. We have drifted away from this way of being, and now we starve for its return.

Turn to life in Nature to restore this essential sense of balance quickly. Allow these lives the space they need within their own boundaries, just as we need space within ours. Embrace the place where we live and extend that same respect to the other lives who share it with us.

When we do, we begin to cultivate fertile ground for trusting relationships within our shared neighborhoods and communities—not as places for us to take from, but as homes where individuals in Nature can also maintain their health, safety, and balance.

Grackle Coming to the Window

Like Redwing, who had been with us all winter, Grackle began coming to our yard on his own, later in the season. He seemed to spend time with Redwings, but now that it was spring, he had Grackle friends—perhaps family—who also came to eat. He is very vocal, much like Redwings. I am certain he gives a high call when the seed cup is empty.

Many prefer to eat seeds from the cup in the bush rather than from the feeder. I feel attached to my Grackle friend. As time goes by, I come to know those who visit our yard. I see them every day, and the same holds true for them. They become familiar with me, too. Some come, never eat, and simply rest near me while I work, leaving when I leave. I do not need to understand this fully; they are welcome, since this area is their home.

One day in early spring, Grackle somehow knew I was in the bedroom. The curtains were open, but the sheers were still closed. Perhaps he could see my silhouette, as I was backlit by the bathroom light. He flew up and landed on the small ledge outside the large window. He moved about, walking back and forth, calling and calling. I went out back for him, and he was sitting on the fence waiting. I filled the cup. When he finished eating and I was back inside, he returned to the window for a brief moment and then left. I am certain it was an acknowledgment that we understood one another.

Petey, Dear Song Sparrow

Another day, Petey, dear Song Sparrow, was looking into the sliding glass door, asking for help because food was needed on a very cold spring morning. It had been freezing the morning before. We had many Birds in our small yard because of the cold, trying to stay warm by eating seeds. Even Robins had come to inspect the seeds. For me, that is when I know conditions are becoming difficult for them.

That same morning, I had placed the two small feeders in their locations, but I had not put out the ground feeder or any additional seeds to help provide warmth. By late morning, I noticed Petey perched on the ledge of the sliding glass door, looking into our window. He was looking for movement—for me.

Birds and Squirrels have learned that they can see inside while perched on the fence in front of the sliding doors. Sometimes they do this, and I know they are wondering or asking for something. At other times, it may simply be curiosity. They may want to know more about my world inside—the unusual place I enter and leave when I move between my home and their world in the yard.

That day, Petey caught my eye, and I gave the minimal wave of my hand that I use to let him know I see him and that it is only me. He continued to look at me and then jumped down to the patio. I looked outside and saw him again. He then flew to the birdbath just to the left.

I knew to open the sliding door, and when I did, he flew a few feet to one of the boulders where I had placed seeds in the past. He stopped there, turned, and looked back at me.

Petey had worked to communicate with me before, so I remained open and diligent in my efforts to "hear" what he—and anyone else—was attempting to convey. This often came through body language, sensations, images, senses, and vocalizations. Squirrels do this all the time. So did Willow, my sweet Dog. I remain respectful of their world and willingly available to offer assistance if they need help.

At that moment, I knew Petey was asking for seeds. I understood immediately through a knowing that came once I heard his expressions in his own way. I changed my shoes and respectfully entered the yard. I put out the seeds and stepped back inside.

As I turned, I watched Petey jump out again from the rain garden and onto a boulder. For a brief moment, he saw the seeds sprinkled in a small amount for those, including himself, who are ground feeders (those who eat from the ground because they find it difficult to perch on feeders). He flew low and fast to the seeds and began to eat.

He needed seeds, just as the others had that morning, and he attempted to let me know. I attempted to understand and succeeded because I wanted to understand…because I had chosen to stop and listen.

We can learn a great deal from others around us by dropping language and listening through all of our other senses. It is truly amazing. When we are in a calm, balanced state and are open to another, words are often unnecessary—even with people. We can simply be while communicating our deepest and most basic needs.

Through the sensations and ways of learning from Nature, about others, and about situations, I have been offered the greatest clarity and the deepest acceptance of forgiveness and kindness. This intelligence has been demonstrated to me—or offered in flashes and through the wisdom of reflection—in very real ways through the energy that radiates from natural systems.

As unusual as this may sound in writing, when I consider it, it does not seem odd at all. Based on my experiences, I am convinced.

Most importantly, I have been shown—through sensations that I feel and know—that unconditional love does extend from opening ourselves to and accepting the pure love present within natural systems. Sometimes it comes as a flash of what we need. Sometimes it is simply shared. Perhaps it is how we feel about ourselves mirrored back to us now that we have evolved.

I have loved and learned from Nature, even when I ignored it. Even when I had no idea.

I had no idea that the same love was reaching for me, too.

Thank you, Petey. In your own special way, thank you for trying with me.

Twenty-Nine
Evolving with Us

Our Thoughts Become Our Future

What we want in our lives must become part of our inner vision, our focus, and our outward expression in the world. Making room for these lives and communities within our own lives requires us to change many thoughts we may have developed, adopted, or inherited. Many of these thoughts are negative and diluted representations of life in Nature. By disregarding these individuals and who they are in their own right, we choke our own need for interconnectedness with them. These patterns of thinking have not evolved in quite some time, but they can.

We are the ones who must take the steps to open the lines of relationship with Nature, and it is we who must examine our intent. Without honest self-awareness—without witnessing where we hurt, where we do not, and what we may need—we are not emotionally prepared to offer clear intent. And without clear intent, there is no fertile ground in which trust can grow with the larger community of life.

As trust begins to bloom outward, perhaps we will start to see one another as the brilliant beings we are and aspire to be. There is no reason why we cannot live respectfully alongside the rest of life on our beautiful and uniquely sacred planet Earth.

Our Oneness with Nature

When we expand our life vision and worldview to include life in Nature—and hold these lives as part of our decisions, knowing, believing, and acting as if there is no separation in our importance, interconnectedly—we step through a doorway into a world that belongs to all of *Us*.

Go to Nature quietly and respectfully, and learn what it means to become deserving of trust again by choosing to be part of this interconnectedness. Know this because it can be felt.

We can become a species known by other species for respecting boundaries—a people who no longer abuse because of who we believe we are or what we feel is our "right" on this planet, but who instead create space for trust and extend that same trust to other life here as well.

We have but one Earth. It is not ours to dispose of as we wish. If we believe we have that right, why are we not acting as stewards?

Ultimately, we can change our ways and invest in life differently—more sustainably and more considerately. We can rein in our needs to what is truly necessary, focusing on what is fundamental, without waste, torture, or destruction. When we begin within our yards, our homes, and our communities, our actions are recognized. Over time, through Nature's support, we become whole. In time, we come to seek out and nurture these relationships—Us, a larger community of life.

If our lives are not working well in some way, could it be because we are not fully present within our own minds, hearts, and lives? And from that place of presence, are we not able to move forward—beginning today—alongside the unique and intelligent Tree, Squirrel, Bird, and Bee, each seeking only to thrive together with the rest of life in Nature? If we want greater balance in our world and in our lives, we must embody that balance first.

We can offer these lives in Nature what they need to meet their needs—always within one another's boundaries—resulting in bonds and deep connections. Through this newfound family, our support community expands to include genuine concern for Nature's well-being, where we want what is best for them in their lives, safely and in health.

We are not so different—life throughout Nature and life throughout our human communities. Through respectful relationships with Nature, we become more whole. This may be what we have been missing, even if we have always loved the world around us. We need these lives within our ethical landscape, held in trust and respect—not for what they give us, but for who they are in their own right, and for who we are in their presence as a healed reflection.

By valuing these lives, we gain the confidence to see, in the mirror of Nature, our own worth. We rediscover our purpose in participating in this interconnected kindness and in honoring our own intelligence and gifts.

And in this mirror, we begin to see one another until the illusion of separation gives way to oneness.

Restoring Balance for All Life

People need Nature. There is a very simple logic behind our basic need for interconnectedness and why we must earn Nature's trust by becoming trustworthy ourselves. We need to make the evolution of our respect for Nature—and for the individuals and natural communities that collectively are Nature—a top priority in our lives.

We need one another.

Any false, human-created bandages and surface solutions will not help in the deep ways we need. Without genuine change, healing, growth, and sustainability will slow, stall, become stagnant, and suffer in ways we have yet to imagine.

Nature touches us softly even when we do not honor interconnectedness because our bodies are drawn to balance. This dormant state of interconnectedness still lives within each of us. Our lives thrive when we nurture these relationships, and we cannot nurture these relationships without also nourishing ourselves.

To help life in Nature, we must want to help these lives for their own sake. We must allow them to live without imposing our hurt or our excessive greed. We must value space rather than assume we know what they want or need. We must treasure emotional and physical boundaries.

We cannot go back and change what has been done, but we can choose differently going forward. We can live in ways that restore balance for all life. We can learn how to work toward this balance by observing the considerate ways other life in the natural world exists. We can correct the poorly designed, fast-and-now systems that have destabilized our world. And we can work together to clean up the harm we have created.

Doing so requires a shift in our perspective—one grounded in trust, positive intention, and respect. This is how healthy relationships function. Nature lives by these principles, whether we recognize it or not. Life throughout Nature values interconnectedness as a way of being. We have rarely taken the time to learn this, and in doing so, we have also neglected the most basic considerations of one another.

We do not need to intrude upon the lives and worlds of others to find solace and comfort. We need to create that sense of peace at home—a place we no longer need to escape in order to feel better. We need communities where we can relax and experience calm and balance close to home and alongside life in Nature, who understand us as well. They are brilliant allies in kindness, if we are willing to let them be.

Nature needs us to recognize our need for interconnectedness and to begin, today, taking clear and intentional steps to become trusted and trustworthy. *Our shared world, our Earth.*

We can begin immediately in our homes, yards, neighborhoods, and communities. Our decisions and actions radiate outward into the world.

Nature needs us to change our mindset, our behavior, and our approach. Now.

These lives are more than willing to support us, teach us, hold space for our emotions, and inspire our aspirations when our motives are aligned with the greater good of Earth. We honor our place within the larger community of life by upholding the inherent value of all living beings. Just as we contribute our talents, skills, and intelligence to the whole, so does life in Nature.

It is time to stop talking and, with an awakened mind and open heart, begin listening and feeling. We must nourish our empathy, for without it we become disconnected from the very interconnectedness we need. Nature can restore our capacity for appreciation, understanding, and emotion—for both happiness and suffering—so that, in the present moment, we do what is best for ourselves because we are doing what is best for life in Nature.

Our need for interconnectedness is like a dim pilot light within each of us. Our desire to live in healthy ways echoes from the natural world of which we are a part. These lives need what we need to survive, and we need what they need. Remembering that these fundamental needs are gifts shared by all life helps guide us.

Our relationships with Nature are among the most important relationships in our lives, reaching to the center of every cell in our bodies. When these relationships are distorted, neglected, exploited, or broken, all of our other relationships are affected—including our relationship with ourselves and with other people.

Our relationships with Nature matter because they matter. We can evolve toward this truth. We feel its absence within us, and we see its absence around us. Life in Nature is worthy of our respect in ways we have only begun to understand, and we can start here and now to rebuild the trust that has been broken.

We make it, or we break it.

Empower yourself.

The Autumn of Willow Winter

How presence can create such a lasting impact in our lives is truly magical. Positive moments and time shared together do create lasting change. We meet people in instances who say or do something that changes our world ever so slightly, sending ripples through the days that follow. A single moment together can transcend time and place, touching our experiences, our senses, our thoughts, our lives, and our worlds. The natural world remains especially fluid in these ways. As we find our way, we begin to understand that these individual lives can become part of Us—together—when we honor their boundaries, rights, and needs just as we honor our own.

Our companion Animals—our pets—are just as special, profound, and brilliant. They share our sacred home and sense of personal space. They often become part of our family if we are wise enough to allow it. We love them, and they love us. They are worth our care and concern and are valued enough that we want a proper life for them, too, beyond our own needs. And over time, with certain individuals, within that very specific and unique sense of love, they become part of our breath. Willow is—was—that with me.

I adopted Willow, sweet Dog, as a baby from a pet store that doubled as an adoption center because people would leave Animals at the storefront door in hopes of giving them a better life (I can only hope). Willow was left with her sister, who had been adopted a few days before. I was told they were going to adopt both sisters, but thought it better to make sure they could do well with one. They planned to come back for Willow. They never did, fortunately, because I held her and knew, after a few weeks of meeting individuals, that she and I—there was something between us. She was tiny and very sick, and I paid a minimal fee that covered some food and care and signed promises to do right by her, be responsible with her, and care for her. She and I went home together for the first time that day.

I named her after Weeping Willows—Trees I love. We grew close, and when she was around two years of age, we became more *one*, and she was a constant in my life. Little did I know how far-reaching and how deeply embedded her roots would become in grounding me.

Each love we have in life is different, even among our sweet pets. Our sense of love is just as unique as the individual.

When she was around thirteen, Willow started to have moments, days, weeks, and for a period of time, months of being up all night. She had started her way into the early stages of decline. There were periods of serious illness. But it was not her time yet. There were more moments of joy and play.

In time, I understood the moments when I needed space. She did, too. I became so determined to prevent the demise of something I knew was coming. For anyone wanting to help someone they love, decline is not always easy. Not all people are as sympathetic to the love we feel for natural life, which includes our pets. I came to understand that I needed to change some things to allow for what I needed so that I could give her what she needed, too. And only this would help her.

As I looked into my sweet Willow's eyes—still puppy-like, even when slightly clouded within a face once black and white and now gray—she would look back and forth into each of my eyes in a way that was different. She did this only a few times. In these moments, without words, something was pulling from within, connecting us through our lives together. This place—this within—connects us, still.

As we walked along community trails and neighborhood streets through our more than seventeen years together, understanding Willow in a way that allowed her to understand me brought us so close that there was no separation. Sacrificing oneself for another completely, even in times when the other is struggling, is not good, and the struggle bleeds over, especially for someone like me. But space with purpose, for the good of the whole, offers the airflow we need to be better in all of our relationships.

My sweet Willow had been with me for so long. All of the most important learning I have been honored to receive and experience, especially the many moments with so many individuals in Nature, Willow was right by my side. Through all of the work and understanding, Willow was my witness.

She struggled for four years, on and off, until it was more struggle and less Willow. Our sense of love together altered in a way that we both needed to adjust to. Our sense of love always remains.

With Willow, interconnectedness includes the knowing that even in death, she remains. The consciousness of our love, play, and joy is unscathed. In the consciousness of interconnectedness, we are closest. We meet through our sense of love.

Willow has taught me so much about grounding myself to what I care about most. With her, this loyalty remains and will for the rest of my days.

Within, Us

Our uniquely individual journey is never undertaken alone. The benefits of nurturing these relationships with Nature are too numerous and too vital to the future of all life. We need time in our yards, outside in the fresh air, alone, quietly, and without distractions that deflect Nature's voice. Begin to want to earn Nature's trust. Make the effort to heal these broken relationships and, in turn, heal within.

If we begin to see other life as part of our life—together—perhaps the separation we have created that has hindered our well-being will begin to fade. For when we see other life as an "it," they will not matter to us in ways that matter to *Us*. Only we can find our future, our health, and our happiness—together. These roots are close to our homes and loved ones, within our self-love, too. Here we might find the human balance many of us have been seeking, and in doing so help restore balance on this planet—the balance we all need. There is no stronger connection to life, to our own Nature, than what we hold in our hearts. Our relationships with these lives around us need to matter to us.

Nature needs people to accept their individually unique intelligence and emotional depth. If we experience it, chances are Plant, Bird, Animal, and Insect life do as well. We are a common family of Earth life, and this was never meant to be taken lightly. In our sense of joy, we will see and know joy in others.

The reality is, we need Nature. As much as some might like to think that we are superior to Nature, we are only part of a much larger picture. They are our sisters and brothers in the natural world. If we wish to do what we can to support Nature, we need to begin to support ourselves, too. In order to save Nature, we need to save our own sense of self. There was never a distinction between self and Nature.

Thirty
Closing Meditation

Respectfully, we can choose to hold the same desire for each individual in the natural world that we hold for ourselves: that they be healthy, safe, and free to thrive within a trusting relationship grounded in active trustworthiness and alive with interconnectedness.

Take a quiet moment to settle into your breath and into the present. Allow yourself to feel your place within the larger community of life. Around you and within you, interconnectedness is already present—alive in every breath, every sensation, and every relationship that supports your existence. As you enter this meditation, consider the possibility that the same desire you hold for your own well-being can be extended to the lives around you. In offering this intention to others, we begin to experience these same gifts within ourselves.

- Focus on demonstrating your true capacity for trustworthiness
- Remain open to information conveyed from Nature through what you sense and feel
- Hold close the understanding that what you *think* you need may not be what you truly need
- Sit respectfully in the backyard, with one Tree in the neighborhood, or alongside Plant in the home
- You cannot force what another will be with you; kind relationships are built through mutual agreement of time, space, and moment
- No spoken words
- Focus on breath
- Remain present within the body as you breathe
- Be calm, present, without tension or anxiety
- Considerately accept another as they are, not as you want them to be
- Quiet
- Learn about yourself through a deeper and more accurate understanding of life in Nature

- Know the individual
- Respect distance, avoid staring, and remain calm, relaxed, and not on edge
- Focus on being present while remaining open to information beyond thought
- When thinking takes over, return focus to the breath, relaxing the body
- Accept individuals and their invaluable role within the equation, just as you may begin to accept your own
- Build an immediate community based on trust and respect
- Acknowledge and appreciate our collective basic need for interconnectedness once again
- Continue building relationships across other forms of natural life
- Information spreads through the nonlocal: nonlocal information can be delayed; flashes of understanding may appear later and unexpectedly
- Who you are within follows you wherever you go
- What is known and understood may arrive later, often unexpectedly
- Trust yourself; these relationships are personal
- Maintain balance and a desire for balance throughout daily life
- Our greater-good passion resides within all of us, and we are responsible for it
- Nature can be your greatest support on Earth if you create the space
- Learn who you are and like who you are
- Want to be better, be open—change, adapt
- Want to embrace interconnectedness with Nature
- Honor boundaries and allow others to be
- Most importantly, understand that you are part of the picture of life on sacred Earth—not merely an observer, but a guest within the collective whole of all life finding home here on this beautiful planet
- There is great wisdom in honoring your role as a guest among many when entering the natural world
- We are what we think, and Nature responds to what we project; words matter
- Embrace the uniqueness of each individual
- Allow others to communicate without assumption or emotional coloring
- Be with Nature throughout the day, even in abbreviated moments—such as when you walk to your car—with the same trust and consideration
- Honor your experiences
- We—*all of us*—are part of Earth, part of the equation
- Remember your inherent worth
- Know Nature's inherent worth
- Trust—create and appreciate the role of trust
- Interconnectedness—our relationships with Nature matter
- Our evolution of respect comes from knowing the individual lives who are Nature
- Gratitude

Acknowledgments

I would like to acknowledge the people who have supported me in bringing this vision and body of work to completion over the years. There have been many.

I want to thank, with great reverence, The Awakened Press team. I extend my gratitude to my publisher and managing editor, Lindsay R.A. Dierking, for feeling through what I needed to relay with words. I also thank the team members who could see the images with little effort, those working to perfect formatting for both print and electronic versions, and the gifted people who helped clarify these experiences without changing my voice. The entire team cares deeply for these brilliant lives who stabilize our very special Earth. Thank you.

Thank you to those who took the time and showed real interest in reading a small part of this work for feedback.

I would like to personally acknowledge the following:

Jenna Benton: With such grace and ease, your insight, words, and commitment to Animals and all Wildlife have been an inspiration. Your words touch so easily. I am eternally grateful for the bright light you have offered me and continue to offer to all life on our Earth.

Mary Kay "Sam" Elsen: Your steady and calm voice of logic has been invaluable. Thank you for your honesty, balanced openness, and for allowing life in your world to reach you beautifully.

William H. Fields, Esq.: You are one who is kind and confident in presence, as I knew in childhood. Even after years, your participation had meaning. Your written messages have made something that can appear poetic more relatable. Thank you for your gift.

Jessie Foley: With such depth, intelligence, and the willingness to learn and grow, while remaining so light and easy to be with, even from a distance, your words and feedback have a powerful energy all of their own. Thank you for feeling.

Karen Naimoli: Many years ago, you took me to our musty, cold basement and taught me how to write my name on an old chalkboard so that I could enter kindergarten. Thank you for caring to take that moment to help me learn something that seems small but was essential.

Thank you for doing the same now.

Michael Naimoli, MSE: In the summer, I sat outside, frustrated. You walked up to me and asked, "What's wrong?" I knew what the end result looked like as I fiddled with my laces. I could not figure out how to twist the loops together from these two sides of the laces. In my own repetitive failed attempts, I said in frustration, "No one will teach me how to tie my shoes." Happily, you said, "I will!" You showed me a few times until I mastered it. Perhaps it was the teacher, but I remember thinking, "That's it?" Thank you for taking the time to focus on what I needed so that I could help myself each day afterward. Thank you for your willingness to help again. Your ability to see something through written words matters.

Brian C. Scott: Your willingness to help, and your kindness, is not unnoticed, as someone who has devoted his life to service extending to the natural world. Thank you for your commitment.

Leslie Whitcomb, Ph.D.: A mental health clinician and ecopsychology educator who glows with warmth, gentle support, and wisdom; thank you for your gift of *breathing* applied ecopsychology. Your honesty and ability to impart clarity are valued and sensed beyond words.

David Winter: I am forever grateful for who you are and what you contribute to the world. Thank you for the balance you provide, including your insight into this work. Your unwavering support of who and what I care about most is like fresh air. The parts of life we share are cherished always—along with our joy and laughter.

I want to thank Mariah for her support and wisdom many years ago. Thank you for hearing me without judgment and for offering a safe space that allowed me to begin sharing my experiences and what I was coming to understand from life in Nature, met only with kindness in return. I have held on to that small spark of kindness ever since—acting as guidance, at times. Your words, "It's okay that you feel this way," changed me forever.

About the Author

Kristen and Willow on their trail.

Kristen Winter, Ph.D., has spent most of her career in nonprofit environmental conservation and sustainability in administration, operations, education, project management, and philanthropy. She has worked with national and international groups, along with local and regional organizations and nature centers. With one foot in the natural world, Kristen has developed a deep appreciation for human individuality. Nature has shown her that profound parallels exist across all life on our uniquely special planet Earth.

With a Bachelor of Science in Psychology and a Master of Science in Environmental Protection and Safety Management, Kristen earned her Ph.D. in Applied Ecopsychology. From there, she expanded her dissertation to more deeply understand her past and Nature's individuals. Guided by profound and empowering moments, Kristen came to realize that these brilliant lives support us in ways we may never have imagined.

With the memory of Willow, her sweet Dog, who remains her grounding force, Kristen and her husband live with a mutual love of finding laughter in the everyday. Together they focus on balance, where they spend personal time away from technology, often just out their door, enjoying the moment, and in the sunshine. This is her first book.